A. M. Yaglom and I. M. Yaglom

Challenging Mathematical Problems with Elementary Solutions

Volume II

Problems From Various Branches of Mathematics

Translated by James McCawley, Jr.

Revised and edited by Basil Gordon

DOVER PUBLICATIONS, INC.

NEW YORK

This Dover edition, first published in 1987, is an unabridged and unaltered republication of the edition published by Holden-Day, Inc., San Francisco, in 1967. It was published then as part of the *Survey of Recent East European Mathematical Literature,* a project conducted by Alfred L. Putnam and Izaak Wirszup, Dept. of Mathematics, The University of Chicago, under a grant from The National Science Foundation. It is reprinted by special arrangement with Holden-Day, Inc., 4432 Telegraph Ave., Oakland, California 94609.

Originally published as *Neelementarnye Zadachi v Elementarnom Izlozhenii* by the Government Printing House for Technical-Theoretical Literature, Moscow, 1954.

Library of Congress Cataloging-in-Publication Data

Yaglom, A. M.

[Neelementarnye zadachi v elementarnom izlozhenii. English]

Challenging mathematical problems with elementary solutions / A. M. Yaglom and I. M. Yaglom : translated by James McCawley, Jr. : revised and edited by Basil Gordon.

p. cm.

Translation of: Neelementarnye zadachi v elementarnom izlozhenii. Reprint. Originally: San Francisco : Holden-Day, 1964–1967.

Bibliography: p.

Includes indexes.

Contents: v. 1. Combinatorial analysis and probability theory—v. 2. Problems from various branches of mathematics.

ISBN-13: 978-0-486-65536-9 (pbk.: v. 1) ISBN-13: 978-0-486-65537-6 (pbk.: v. 2)
ISBN-10: 0-486-65536-9 (pbk.: v. 1) ISBN-10: 0-486-65537-7 (pbk.: v. 2)

1. Combinatorial analysis—Problems, exercises, etc. 2. Probabilities—Problems, exercises, etc. 3. Mathematics—Problems, exercises, etc. I. Yaglom, I. M. (Isaac Moiseevich), 1921– II. Gordon, Basil. III. Title.

QA164.I1613 1987
511′.6—dc19 87-27298
CIP

Manufactured in the United States by LSC Communications
65537712 2022
www.doverpublications.com

PREFACE TO THE AMERICAN EDITION

As in the case of Volume I, a considerable number of changes have been made in adapting the original book for use by English-speaking readers. The editor accepts full responsibility for these changes which include the following:

1. The order of a few problems has been changed in order that no problem need depend on a later one for its solution.
2. Problems 110 and 128 were added in order to supplement and lend perspective to problems 112 and 127, respectively.
3. Several solutions, and some of the introductory paragraphs, have been expanded or rewritten to bring out points not familiar to many American readers. This applies in particular to most of Section 8.
4. The bibliography has been considerably enlarged.

BASIL GORDON

Cambridge
1967

SUGGESTIONS FOR USING THE BOOK

This volume contains seventy-four problems. The statements of the problems are given first, followed by a section giving complete solutions. Answers and hints are given at the end of the book. For most of the problems the reader is advised to find a solution by himself. After solving the problem, he should check his answer against the one given in the book. If the answers do not coincide, he should try to find his error; if they do, he should compare his solution with the one given in the solutions section. If he does not succeed in solving the problem alone, he should consult the hints in the back of the book (or the answer, which may also help him to arrive at a correct solution). If this is still no help, he should turn to the solution. It should be emphasized that an attempt at solving the problem is of great value even if it is unsuccessful: it helps the reader to penetrate to the essence of the problem and its difficulties, and thus to understand and to appreciate better the solution presented in the book.

But this is not the best way to proceed in all cases. The book contains many difficult problems, which are marked, according to their difficulty, by one, two, or three asterisks. Problems marked with two or three asterisks are often noteworthy achievements of outstanding mathematicians, and the reader can scarcely be expected to find their solutions entirely on his own. It is advisable, therefore, to turn straight to the hints in the case of the harder problems; even with their help a solution will, as a rule, present considerable difficulties.

The book can be regarded not only as a problem book, but also as a collection of mathematical propositions, on the whole more complex than those assembled in Hugo Steinhaus's excellent book, *Mathematical Snapshots* (New York: Oxford University Press, 1960), and presented in the form of problems together with detailed solutions. If the book is used in this way, the solution to a problem may be read after its statement is clearly understood. Some parts of the book, in fact, are so written that this is the best way to approach them. Such, for example, are problems 125 and 170, and, in general, all problems marked with three asterisks.

The problems are most naturally solved in the order in which they occur. But the reader can safely omit a section he does not find interesting. There is, of course, no need to solve all the problems in one section before passing to the next.

This book can well be used as a text for a school or undergraduate mathematics club. In this case the additional literature cited in the text will be of value. While the easier problems could be solved by the participants alone, the harder ones should be regarded as "theory." Their solutions might be studied from the book and expounded at the meetings of the club.

CONTENTS

CHALLENGING MATHEMATICAL PROBLEMS WITH ELEMENTARY SOLUTIONS

Volume II

Problems From Various Branches of Mathematics

PROBLEMS

I. POINTS AND LINES

Problems 101 to 107 are taken from a branch of mathematics called *projective geometry*. We are concerned here only with *plane* projective geometry, which deals with the properties of plane figures that are unchanged by projection from one plane onto another (fig. 1).

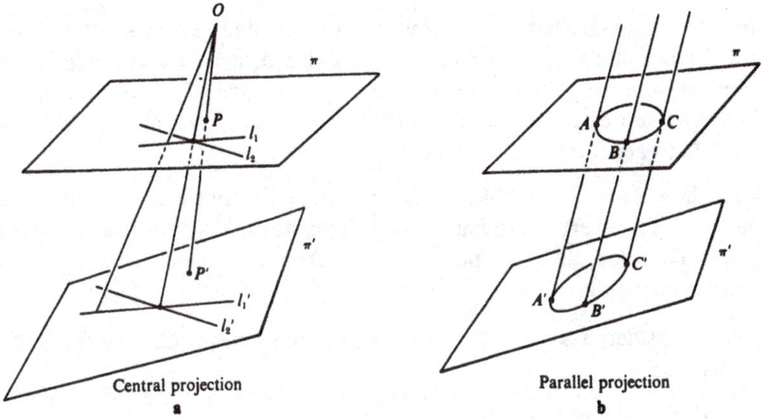

Fig. 1

There are two kinds of projection, *central* and *parallel*. A central projection is obtained by choosing two planes π, π', and a center O which is on neither of them. Given any point P of π, we draw OP and extend it until it intersects π' in a point P' as in fig. 1a. The mapping $P \to P'$ is called a central projection. (Of course if OP is parallel to π' there will be no point P' of intersection. In this case P can be thought of as being mapped to a "point at infinity" by the projection.) A parallel projection from π to π' is obtained by drawing a family of parallel lines and mapping each point A of π onto the point A' of π' such that AA' is one of these lines (fig. 1b). This is sometimes thought of as central projection from a point at infinity.

A projection sends any figure F drawn in the plane π to a figure F' in π'. We call F' the *image* of F. The image of a straight line is itself a straight line, but the distance from a point P to a line l, or the angle between two lines l_1 and l_2, may be changed by projection (fig. 1a). Similarly, the image of a circle need not be a circle (fig. 1b). Projective geometry deals only with the properties of figures which are unchanged by projections, and is therefore not concerned with such things as distances, angles, and circles. However, concepts involving only the incidence of points and lines (such as collinearity or concurrence) are preserved under projection and therefore belong to projective geometry. Each configuration of straight lines and curves in problems 101 to 107 would be transformed under projection into another configuration having the same properties, and thus the results proved in these problems are theorems of projective geometry.

For further reading see [4] and [10].

101. A certain city has 10 bus routes. Is it possible to arrange the routes and the bus stops so that if one route is closed, it is still possible to get from any one stop to any other (possibly changing along the way), but if any two routes are closed, there are at least two stops such that it is impossible to get from one to the other?

102. Show that it is possible to set up a system of bus routes (more than one) such that every route has exactly three stops, any two routes have a stop in common, and it is possible to get from any one stop to any other without changing.

103.* Consider a system of at least two bus lines with the following properties:

(1) Every line has at least three stops.
(2) Given any two stops there is at least one bus line joining them.
(3) Any two distinct lines have exactly one stop in common.

a. Show that all the lines have the same number of stops. Calling this number $n + 1$, show that every stop lies on $n + 1$ different lines.

b. Prove that there are altogether $n^2 + n + 1$ stops and $n^2 + n + 1$ lines in the system.

104a. Arrange nine points and nine straight lines in the plane in such a way that exactly three lines pass through each point, and exactly three points lie on each line.

b. Show that such an arrangement is impossible with seven points and seven straight lines.

105.* Let S be a finite set of straight lines in the plane, arranged in such a way that through the point of intersection of any two lines of S there passes a third line of S. Prove that the lines of S are either all parallel or all concurrent.

106.* Let S be a finite set of points in the plane, arranged in such a way that the line joining any two points of S contains a third point of S. Prove that all the points of S are collinear.

Remark. The results of problems 105 and 106 are "dual" to each other in the sense of projective geometry. This means that either result is obtained from the other by interchanging the words *point* and *line* and interchanging the terms *line joining two points* and *point of intersection of two lines.* The "principle of duality" asserts that whenever a theorem holds in projective geometry, so does its dual. See, for example, pp. 13–14 in [4].

107.** In the plane we are given n points, not all collinear. Show that at least n straight lines are required to join all possible pairs of points.

108a. Find all possible arrangements of four points in the plane such that the distance between any two of them is one or the other of two given quantities a and b. Find all the values of the ratio $a:b$ for which such arrangements are possible.

b. Find all possible arrangements of n points in the plane such that the distance between any two of them is either a or b. For what values of n do such arrangements exist?

109a.* Show that for any integer $N > 2$ it is possible to find N points in the plane, not all on one line, such that the distance between any two of them is an integer.

b.** Show that it is impossible to find infinitely many points in the plane satisfying the conditions of part **a**.

II. LATTICES OF POINTS IN THE PLANE

Problems 110 to 112 deal with lattice points in the plane, that is, with a system of points at the vertices of a network of squares (the *lattice squares*) covering the plane in the same way as the squares on a sheet of graph paper cover the sheet (fig. 2). Such lattices play an important part in pure mathematics (theory of numbers) and in scientific applications (crystallography): Minkowski's theorem (problem 112) is particularly important and has many applications in the theory of numbers. See Ref. [19], chap. IV and [10], chap. II.

In some of the following problems it is convenient to introduce a coordinate system, as in fig. 2, and to take the width of the lattice squares

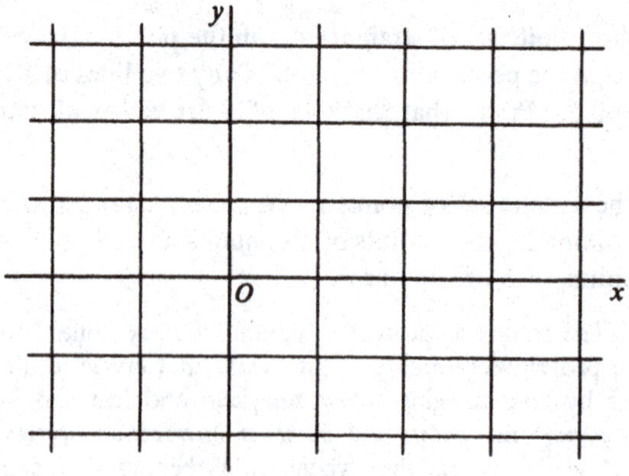

Fig. 2

as the unit of measurement. Then the lattice points are precisely the points (x,y) whose coordinates x and y are integers.

110. *Blichfeldt's lemma.*[0] Let M be a set in the plane with area greater than 1. Show that M contains two distinct points (x_1,y_1) and (x_2,y_2) such that $x_2 - x_1$ and $y_2 - y_1$ are integers.

111a. Let Π be a parallelogram whose vertices are lattice points, and suppose there are no other lattice points inside Π or on its boundary (fig. 3a). Prove that the area of Π is equal to that of a lattice square.

b. Let Π be any polygon whose vertices are lattice points (fig. 3b). Prove that the area A of Π is given by the formula $A = i + b/2 - 1$, where i is the number of lattice points inside Π and b is the number of lattice points on the boundary of Π. Here the unit of area is that of a lattice square. For example, the area of the polygon in fig. 3b is $4 + \frac{16}{2} - 1 = 11$.

(Part **a** is a special case of part **b**, with $i = 0$ and $b = 4$.)

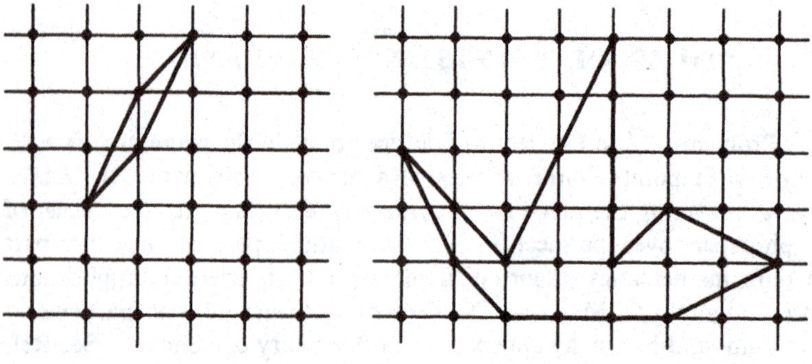

a b

Fig. 3

[0] Hans Blichfeldt (1873–1945), an American mathematician.

112.** *Minkowski's theorem.*[1] Let K be a convex set in the plane which is centrally symmetric with respect to the origin O. Suppose K has an area greater than 4. Prove that K contains a lattice point other than O.

113. In a circular orchard of radius 50 with center at the origin O, trees are planted at all the lattice points except O. In the middle of the orchard there is a garden-house (fig. 4). As long as the trees (which we assume have equal circular cross section) are thin enough, they do not block off the

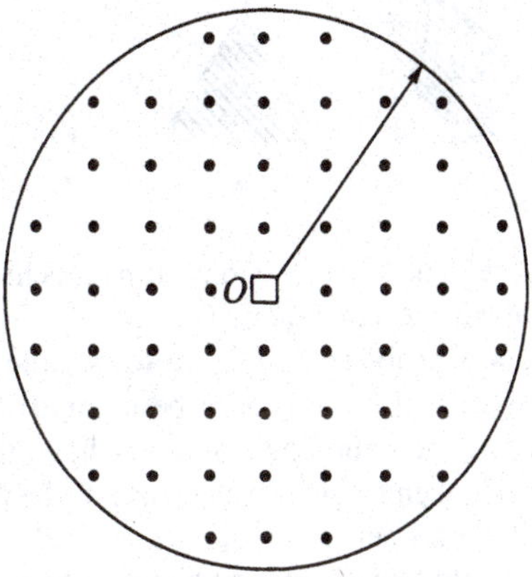

Fig. 4

view from the garden-house (that is, it is possible to draw a ray from O which does not pass through any of the trees). Show that it is possible to see out of the orchard when the trees have radius less than $1/\sqrt{2501} \approx 1/50.01$, but that when the radius becomes greater than $1/50$ the view is completely blocked off.

III. TOPOLOGY

Topology is a branch of mathematics which is concerned with very general, purely qualitative properties of geometric figures. As a branch of mathematics in its own right, it arose comparatively recently, in the twentieth century. At present it is an important part of mathematics, having valuable applications to many other branches. See Refs. [2], [7], and [10], chap. VI.

114. Draw n straight lines in the plane. Show that the regions into which these lines divide the plane can be colored with two colors in such a way

[1] Hermann Minkowski (1864–1909), a German mathematician.

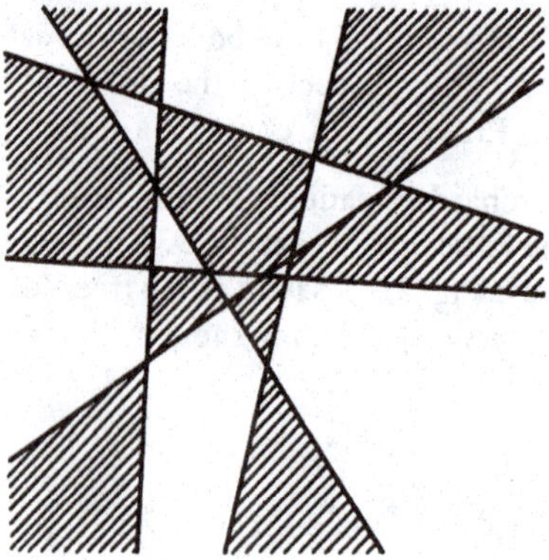

Fig. 5

that no two adjacent regions (that is, two regions touching along a segment of one of the lines) have the same color (fig. 5).

We may formulate problem 114 as follows: the map obtained by drawing n straight lines in the plane can be colored using two colors in such a way that no two neighboring countries have the same color. In this form the problem is seen to be a special case of the following question: What is the minimum number of colors which will suffice to color *any* map in such a way that no two neighboring countries have the same color? This question is still unanswered, although it has attracted the attention of mathematicians for the past one hundred years. It has been proved that five colors are enough, and it is conjectured that four colors are enough. Problems on the coloring of networks of lines (see 115) and nodes (see Remark to 117) are also connected with this question. For more details, see reference [7].

115a. Consider a network of lines and nodes with the property that at most two lines meet at each node (fig. 6a). The lines are to be colored in such a way that no two adjacent lines (that is, lines meeting at a node) have

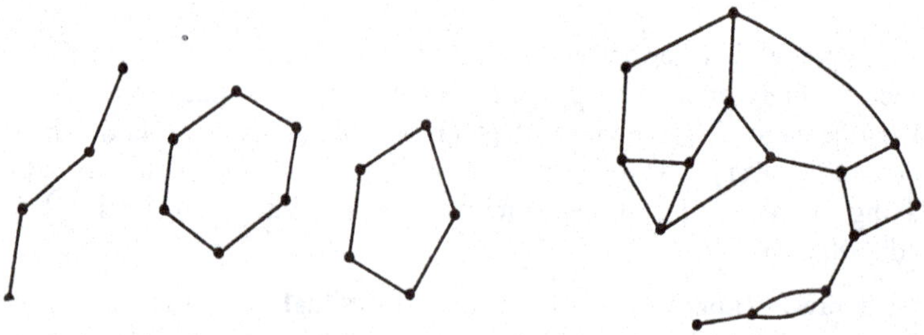

Fig. 6

the same color. Show that this is always possible using three colors, but may be impossible with only two colors.

b.** Suppose that at most three lines meet at each node of the network (fig. 6b). Show that the lines can be colored as in part **a** using four colors, but that this may be impossible with only three colors.

Remark. In practice it sometimes happens that a network of lines has to be colored in the way described above. This occurs, for example, when an electrical network has to be connected without mixing up the wires. It proves convenient to use wires of different colors and never to plug two wires of the same color into the same terminal.

116a.** *Sperner's lemma.*[2] A triangle T is divided into smaller triangles in such a way that any two of the small triangles have no point in common,

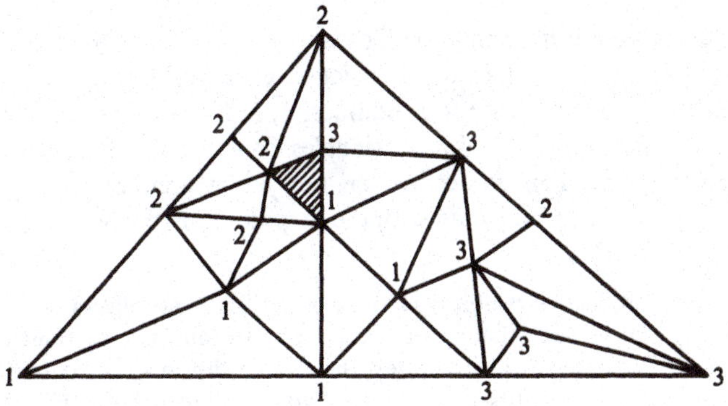

Fig. 7

or have a vertex in common, or have a complete side in common. (Thus no two triangles touch along *part* of a side of one of them.) The three vertices of T are numbered 1, 2, 3. Each vertex of the small triangles is then also numbered 1, 2, or 3. The numbering is arbitrary except that vertices lying on the side of T opposite vertex i (for $i = 1, 2, 3$) must not be numbered i (fig. 7). Show that among the small triangles there is at least one whose vertices are numbered 1, 2, 3.

b. State and prove an analogous result for a tetrahedron divided into smaller tetrahedra.

117.* A triangle is divided into smaller triangles as in problem 116**a**. Show that if an even number of triangles meet at each vertex, then the vertices can be numbered 1, 2, or 3 in such a way that the vertices of every triangle have three different numbers attached to them (fig. 8).

[2] Emmanuel Sperner (1905–), a German mathematician.

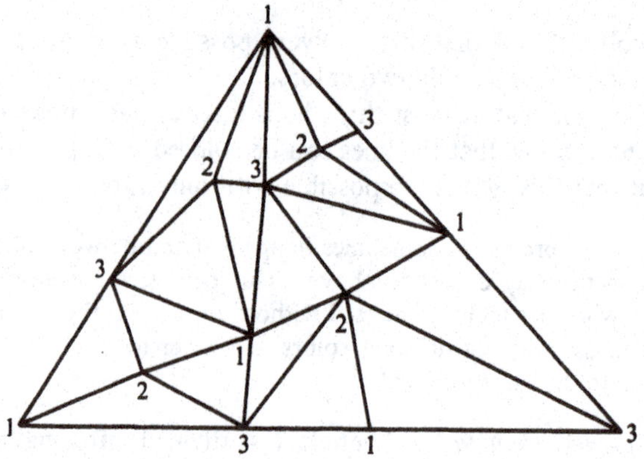

Fig. 8

There is a conjecture that in all cases (that is, without the requirement that an even number of triangles meet at each vertex), it is possible to number the vertices with the four numbers 1, 2, 3, and 4 in such a way that the vertices of any of the small triangles have three different numbers attached to them. This conjecture has not yet been either proved or disproved. It is closely connected with the map-coloring problem (see above, under problem 114).

Remark. Note that problem 117 may be rephrased as follows: if a triangle is broken up into small triangles, as described in problem 116a, and an even number of triangles meet at each vertex, then the vertices may be colored so that no two adjacent vertices (that is, vertices belonging to the same triangle) have the same color, using just three colors (see problems 114 and 115).

118.*** *A problem on neighbors.* A square of side 1 is divided into polygons (fig. 9). Suppose that each of these polygons has a diameter[3] less than $\frac{1}{30}$. Show that there is a polygon P with at least six neighbors, that is, polygons touching P in at least one point.

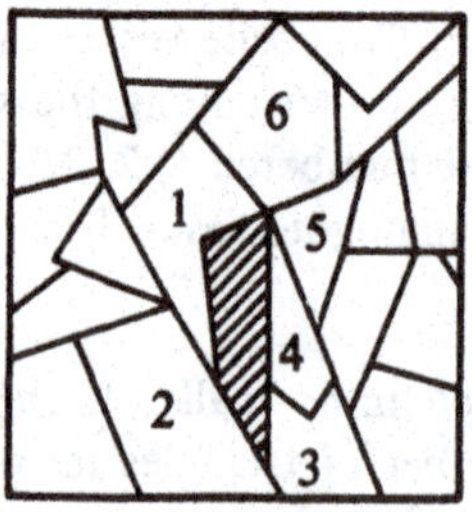

Fig. 9

[3] By the *diameter* of a polygon we mean the greatest distance between any two of its points.

IV. A PROPERTY OF THE RECIPROCALS OF INTEGERS

119. Let $a = 1/n$ be the reciprocal of a positive integer n. Let A and B be two points of the plane such that the segment AB has length 1. Prove that every continuous curve joining A to B has a chord[4] parallel to AB and of length a. Show that if a is not the reciprocal of an integer, then there is a continuous curve joining A to B which has no such chord of length a.

For a complete characterization of the possible sets of chord lengths parallel to AB, see H. Hopf [26].

V. CONVEX POLYGONS

Problems 120 to 122 are concerned with extremal properties of convex polygons. The results are true, and the proofs need little alteration, for any convex sets (not necessarily polygons). A set is said to be convex if the whole of the line segment joining any two points in it lies inside the body.

The theory of convex bodies is an important part of geometry, having applications in other branches of mathematics and science. See Refs. [19], [22], and [24].

120a. Show that any convex polygon of area 1 can be enclosed in a parallelogram of area 2.

b. Show that a triangle of area 1 cannot be enclosed in a parallelogram of area less than 2.

121a. Show that any convex polygon of area 1 can be enclosed in a triangle of area 2.

b.** Show that a parallelogram of area 1 cannot be enclosed in a triangle of area less than 2.

122a.* Let M be a convex polygon and l any straight line. Show that it is possible to inscribe a triangle in M, one of whose sides is parallel to l, and of area at least $\frac{3}{8}$ that of M.

b. Let M be a regular hexagon, and l a line parallel to one of its sides. Show that it is impossible to inscribe a triangle in M with one side parallel to l and of area more than $\frac{3}{8}$ that of M.

[4] A *chord* of a curve C is a straight line segment, both of whose endpoints lie on C.

VI. SOME PROPERTIES OF SEQUENCES OF INTEGERS

123a.** Consider n arithmetic progressions, each consisting entirely of integers, and extending indefinitely in both directions. Show that if any two of the progressions have a term in common, then they all have a term in common. Show that if the progressions are allowed to assume non-integer values, then the conclusion may be false.

b. Consider n arithmetic progressions, each extending indefinitely in both directions. If each three of them have a term in common, prove that they all have a term in common.

Remark. It is interesting to note the similarity between the phrasing of this problem and the following theorem of Helly: If a family of convex sets in the plane is such that any three have a point in common, then all of them have a point in common. For a unified treatment of both theorems, see reference [27].

124a. Show that a sequence of 1's and 2's, having at least four terms, must contain a digit or sequence of digits which appears twice in succession.

b.** Show that there exist arbitrarily long sequences of 1's and 2's in which no digit or sequence of digits occurs three times in succession. (Cf. M. Morse and G. A. Hedlund, "Symbolic Dynamics", American Journal of Mathematics, vol. 60 (1938), pp. 815–866.)

125a.*** Show that there exist arbitrarily long sequences consisting of the digits 0, 1, 2, 3, such that no digit or sequence of digits occurs twice in succession.

b. Show that there are solutions to part **a** in which the digit 0 does not occur. Thus three digits is the minimum we need to construct sequences of the desired type.

Although part **a** is actually a corollary of part **b**, we have stated the problems separately, for part **a** is somewhat simpler.

126.** Let T be a positive integer whose digits consist of N 0's and 1's. Consider all the n-digit numbers (where $n < N$) obtained by writing down n consecutive digits of T; there are $N - n + 1$ such numbers, starting at the 1st, the 2nd, . . . , the $(N - n + 1)$th digit of T.[5]

Show that, given n, we may choose N and T in such a way that all the n-digit numbers obtained from T are different, and that *every* n-digit number consisting entirely of 1's and 0's is to be found among them.

[5] Thus if $T = 10010$, then the two-digit numbers we obtain in this way are 10, 00, 01, and 10 again.

VII. DISTRIBUTION OF OBJECTS

127. We are given 10 cookies of each of 20 flavors. The cookies are distributed into 20 boxes, 10 cookies to each box. Show that however the distribution is made, it is always possible to select one cookie from each box in such a way that the 20 cookies so obtained are all of different flavors.

128. *The marriage problem.* Suppose that there are m boys and M girls, and that each boy is acquainted with a certain number of the girls. Suppose, moreover, that for each subset of k boys ($1 \leqq k \leqq m$), the total number of their acquaintances is at least k. Show that it is possible for each boy to marry one of his acquaintances without bigamy being committed.

VIII. NONDECIMAL COUNTING

The problems in this section are related by the fact that their solution involves the use of number systems other than the familiar decimal system. For the reader's convenience, we now give a brief account of these systems.

Consider an infinite sequence of integers $u_0 = 1, u_1, u_2, \ldots, u_n, \ldots$, with $u_0 < u_1 < u_2 < \cdots$. Let N be any given positive integer, and suppose u_n is the largest member of the sequence which is $\leqq N$. We divide N by u_n, obtaining a quotient q_n and a remainder r_n; that is, $N = q_nu_n + r_n$, where $q_n = [N/u_n]$ and $0 \leqq r_n < u_n$. Next we divide r_n by u_{n-1}, obtaining a quotient q_{n-1} and a remainder r_{n-1}; thus $r_n = q_{n-1}u_{n-1} + r_{n-1}$, where $q_{n-1} = [r_n/u_{n-1}]$, and $0 \leqq r_{n-1} < u_{n-1}$. We now divide r_{n-1} by u_{n-2}, and proceed in this way, obtaining the equations

$$\begin{array}{rlr} N = q_nu_n + r_n & 0 \leqq r_n < u_n & \\ r_n = q_{n-1}u_{n-1} + r_{n-1} & 0 \leqq r_{n-1} < u_{n-1} & \\ r_{n-1} = q_{n-2}u_{n-2} + r_{n-2} & 0 \leqq r_{n-2} < u_{n-2} & \\ \vdots & \vdots & (1) \\ r_{i+1} = q_iu_i + r_i & 0 \leqq r_i < u_i & \\ \vdots & \vdots & \\ r_2 = q_1u_1 + r_1 & 0 \leqq r_1 < u_1 & \\ r_1 = q_0u_0 & & \end{array}$$

The last remainder r_0 is zero, since $u_0 = 1$. It is possible that some earlier remainder r_i is zero, in which case $q_{i-1} = q_{i-2} = \cdots = q_0 = 0$.

Combining equations (1) we get

$$N = q_n u_n + q_{n-1}u_{n-1} + \cdots + q_1 u_1 + q_0, \tag{2}$$

where $q_i = [r_{i+1}/u_i]$. Since $0 \leqq r_{i+1} < u_{i+1}$, we have $0 \leqq q_i < u_{i+1}/u_i$. We will refer to equation (2) as the representation of N in the number system determined by $\{u_0, u_1, u_2, \ldots\}$. Note that $r_i = q_{i-1}u_{i-1} + \cdots + q_1 u_1 + q_0$, and therefore, since $r_i < u_i$, we have

$$q_{i-1}u_{i-1} + \cdots + q_1 u_1 + q_0 u_0 < u_i \qquad (1 \leqq i \leqq n + 1). \tag{3}$$

It is easily seen that, conversely, given a sequence $q_n, q_{n-1}, \ldots, q_1, q_0$ of nonnegative integers satisfying condition (3), the expression $q_n u_n + q_{n-1}u_{n-1} + \cdots + q_1 u_1 + q_0 u_0$ is the representation of its sum in the number system determined by $\{u_0, u_1, u_2, \ldots\}$. The inequalities $0 \leqq q_i < u_{i+1}/u_i$ are not always sufficient for this, as we shall see later (page 15).

A case of particular importance arises if we choose a fixed integer $b > 1$ and let $u_i = b^i$. Then $u_{i+1}/u_i = b^{i+1}/b^i = b$, and our expansion takes the form

$$N = q_n b^n + q_{n-1}b^{n-1} + \cdots + q_1 b + q_0, \tag{4}$$

where $0 \leqq q_i < b$. When $b = 10$, this means that $0 \leqq q_i \leqq 9$. In this case $q_n, q_{n-1}, \ldots, q_0$ are the digits of N in the usual decimal notation, which is customarily abbreviated as $N = q_n q_{n-1} \cdots q_0$ (not a product!). In general, equation (4) is referred to as the representation of N to the base b, and the resulting number system is called the b-ary system. (For $b = 2$, 3, and 10 the terms *binary* system, *ternary* system, and *decimal* system are preferred.)

It is easy to show conversely that given integers $q_n, q_{n-1}, \ldots, q_1, q_0$, with $0 \leqq q_i < b$, the expression $q_n b^n + q_{n-1}b^{n-1} + \cdots + q_1 b + q_0$ is the representation of its sum to the base b; as noted earlier, this property need not hold for more general sequences $\{u_0, u_1, u_2, \ldots\}$.

The case $b = 2$, that is, the binary system, is especially important and is used in solving problems 129 and 130. Here the "digits" q_i satisfy $0 \leqq q_i < 2$, so that each q_i is either 0 or 1. Hence the representation

$$N = q_n 2^n + q_{n-1}2^{n-1} + \cdots + q_1 2 + q_0$$

gives us an expression of N as a sum of distinct powers of 2, namely, those powers 2^i for which $q_i = 1$. For example, the binary expansions of the integers $\leqq 10$ are as follows:

		Abbreviated notation
$1 =$	2^0	1
$2 =$	2^1	10
$3 =$	$2^1 + 2^0$	11
$4 =$	2^2	100
$5 =$	$2^2 + 2^0$	101
$6 =$	$2^2 + 2^1$	110
$7 =$	$2^2 + 2^1 + 2^0$	111
$8 =$	2^3	1000
$9 =$	$2^3 + 2^0$	1001
$10 =$	$2^3 + 2^1$	1010

In the right-hand column we show the abbreviated notation, in which only the digits $q_nq_{n-1}\cdots q_1q_0$ are written, just as in the decimal system.

As another example (which is of use in problem 131), we consider the case where the u_i are the *Fibonacci numbers*, that is, $u_0 = 1$, $u_1 = 2$, and $u_i = u_{i-1} + u_{i-2}$ for $i \geqq 2$. The first few of these numbers are shown in the following table.

i	0	1	2	3	4	5	6	7	8	9	10
u_i	1	2	3	5	8	13	21	34	55	89	144

The number system determined by this sequence is called the *Fibonacci system*, or F-system for short. Since $0 \leqq q_i < u_{i+1}/u_i = (u_i + u_{i-1})/u_i < (u_i + u_i)/u_i = 2$, we see that each q_i is either 0 or 1, as in the case of the binary system. But it is *not* true that every expression of the form $q_nu_n + \cdots + q_0u_0$, with $0 \leqq q_i \leqq 1$, is the representation of its sum in the F-system. In fact, we cannot have two consecutive digits q_{i-1} and q_{i-2} equal to 1. For if $q_{i-1} = q_{i-2} = 1$, then $q_{i-1}u_{i-1} + q_{i-2}u_{i-2} + \cdots + q_0u_0 \geqq u_{i-1} + u_{i-2} = u_i$, a contradiction to inequality (3) above.

On the other hand, if $1 = q_n, q_{n-1}, \ldots, q_1, q_0$ is a sequence of 1's and 0's, such that no two consecutive terms are 1, then the expression $q_nu_n + q_{n-1}u_{n-1} + \cdots + q_1u_1 + q_0u_0$ is the F-expansion of its sum N. For if $1 \leqq i \leqq n + 1$, we have

$$q_{i-1}u_{i-1} + \cdots + q_1u_1 + q_0u_0 \leqq u_{i-1} + u_{i-3} + u_{i-5} + \cdots u_k,$$

where $k = 0$ or 1 according as i is odd or even. If i is odd, then

$$\begin{aligned} &u_{i-1} + u_{i-3} + u_{i-5} + \cdots + u_2 + u_0 \\ &\quad = (u_i - u_{i-2}) + (u_{i-2} - u_{i-4}) + (u_{i-4} - u_{i-6}) + \cdots + (u_3 - u_1) + u_0 \\ &\quad = u_i - u_1 + u_0 = u_i - 1. \end{aligned} \tag{5}$$

If i is even, then

$$\begin{aligned} &u_{i-1} + u_{i-3} + u_{i-5} + \cdots + u_3 + u_1 \\ &\quad = (u_i - u_{i-2}) + (u_{i-2} - u_{i-4}) + \cdots + (u_2 - u_0) \\ &\quad = u_i - u_0 = u_i - 1. \end{aligned} \tag{6}$$

Thus in any case,

$$q_{i-1}u_{i-1} + \cdots + q_0u_0 \leqq u_i - 1 < u_i,$$

so that the inequalities (3) are satisfied. This shows that $q_nu_n + \cdots + q_0u_0$ is the F-expansion of its sum.

The F-expansions of the integers from 1 to 10 are as follows:

	Abbreviated form
1 = 1	1
2 = 2	10
3 = 3	100
4 = 3 + 1	101
5 = 5	1000
6 = 5 + 1	1001
7 = 5 + 2	1010
8 = 8	10000
9 = 8 + 1	10001
10 = 8 + 2	10010

In problems 130 and 131 we are concerned with mathematical games of the following type.

A network of nodes and lines is given, the one illustrated in fig. 10 for example. There is a unique highest node, marked START. The game is played by two players with a counter which is initially placed at START. The first player moves the counter downward along one of the lines emanating from START, until it covers an adjacent node. For example, in fig. 10 the first player could move to any one of the positions marked 1, 2, or 3. Next, the second player moves the counter farther down in the same manner. In the figure, if the first move is to position 1, the second move must be to 4; if the first move is to 2, the second move must be to

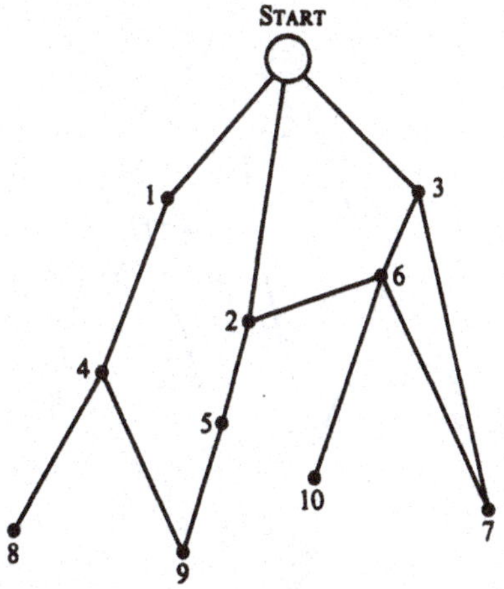

Fig. 10

5; while if the first move is to 3, the second move could be to either 6 or 7. The game continues in this way, with the two players alternately moving the counter downward, until a position is reached where no further move is possible (nodes 7, 8, 9, and 10 in the figure). The player whose turn it is to move is then unable to play, and loses the game. In other words, the object is to make the last move.

We wish to describe briefly how such games can be analyzed. By a *winning position* we mean one with the property that whoever plays from it can win the game by proper play, no matter what his opponent does. The other positions are called *losing positions*; they are such that whoever has to move from them will be defeated by a skilled opponent. The following three properties are clear:

(1) The nodes at the bottom of the network are losing positions.
(2) A node A is a winning position if it is possible to move from A to a losing position.
(3) A node B is a losing position if every move from B leads to a winning position.

By using these three properties we can determine the nature of every node in the diagram, starting at the bottom and working up. For example, in fig. 10 we first mark the bottom nodes with an L (for *losing*). This gives fig. 11.

Next, using property (2), we see the nodes 3, 4, 5, and 6 are winning positions; marking them with a W, we obtain fig. 12. By property (3) it now follows that nodes 1 and 2 are losing positions; this gives fig. 13.

Finally, we see from property (2) that the START is a winning position. This means that the first player can win the game by moving either to 1 or 2, and then, after his opponent moves, pushing the counter to the bottom.

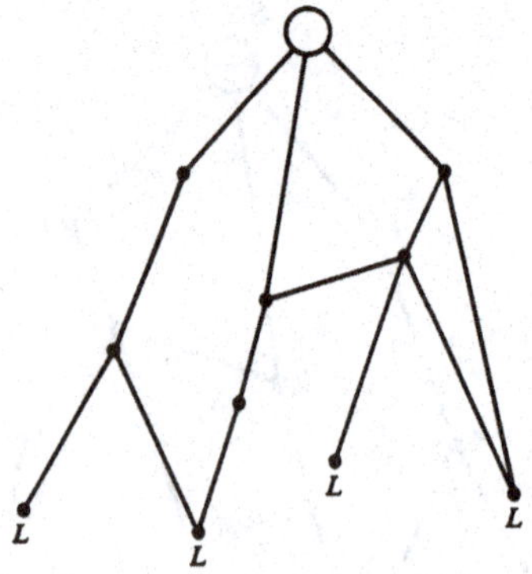

Fig. 11

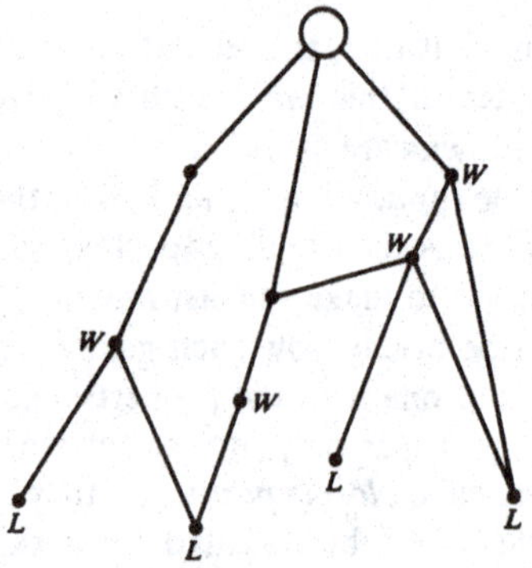

Fig. 12

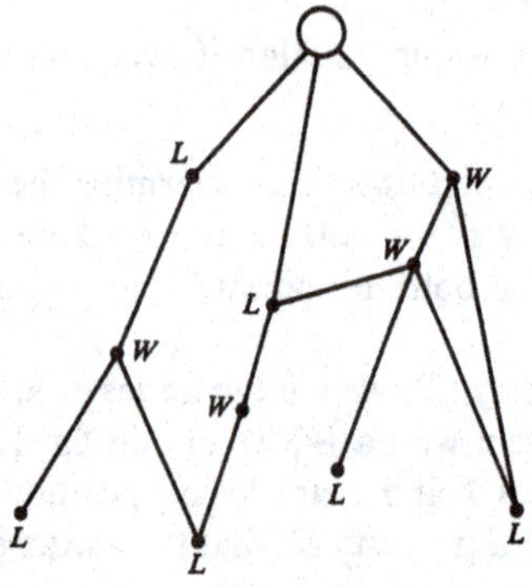

Fig. 13

It is clear from our analysis that with proper play on both sides, the game is a win for the first or second player according as the START is marked with a *W* or an *L*.

In actual practice, the method we have just described for finding the winning and losing positions can be quite cumbersome to apply. The following theorem gives a convenient criterion for identifying the set of *all* losing positions in the network without working up from the bottom.

Theorem. Let $\mathfrak{L}$ be a subset of the nodes of the network such that

(4) No two nodes of $\mathfrak{L}$ are connected by a line of the network.

(5) From every node not in $\mathfrak{L}$ there is a line leading (downward) to a node of $\mathfrak{L}$.

Then $\mathfrak{L}$ is precisely the set of all losing positions.

Proof. The nodes at the bottom of the network are in $\mathfrak{L}$, for otherwise property (5) would be violated. Now suppose you are playing the game and find yourself at a node of $\mathfrak{L}$. By (4) you must move the counter to a node not in $\mathfrak{L}$. By (5) your opponent can then move it back to a node of $\mathfrak{L}$. Continuing in this way, you will find yourself at lower and lower nodes of $\mathfrak{L}$, until you finally reach the bottom of the network; you will then be unable to move. Thus the nodes of $\mathfrak{L}$ are losing positions. On the other hand, if you start at a node not in $\mathfrak{L}$, you can move the counter to a node in $\mathfrak{L}$ by (5). Then you have put your opponent in a losing position and thus have won the game. Hence the nodes not in $\mathfrak{L}$ are winning positions.

The use of this theorem will become apparent in problems 130 and 131.

129.*** The squares of an infinite chessboard are numbered successively as follows: in the corner we put 0, and then in every other square we put the smallest nonnegative integer that does not appear to its left in the same row or below it in the same column (fig. 14). What number will appear at the intersection of the 1000th row and the 100th column?

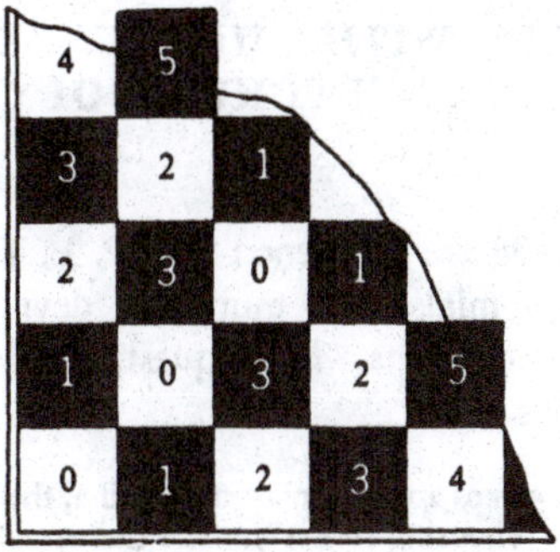

Fig. 14

130.** *The game of Nim.* Two players play with three piles of matches. Each player in turn takes a number of matches (as many as he likes) from any of the piles (but only from one pile in each turn). The winner is the one who takes the last match.

Determine the initial conditions for which the first player can force a win, and those for which he cannot, and give a method of play with which he will always win in the first case.

Remark. Instead of three piles of matches we may use a piece of paper with three rows of squares and three checkers (fig. 15). The players alternately

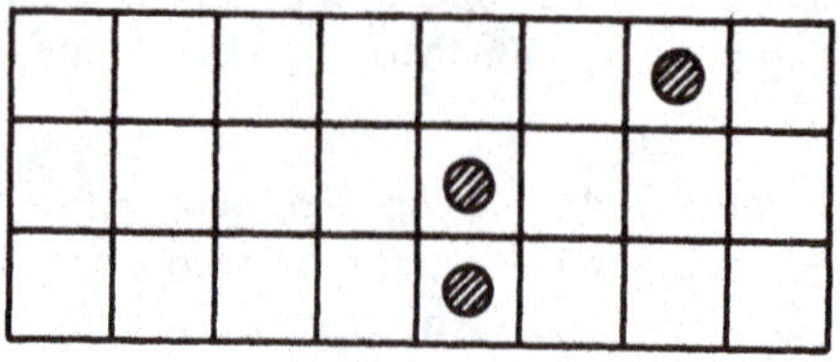

Fig. 15

move one of the checkers any number of squares to the right, and the winner is the one who makes the last move. In this form the game may conveniently be played in a classroom, using chalk and an eraser.

Similar remarks apply to the following problem.

131.*** *Wythoff's game.*[6] The game is played with two piles of matches. The two players alternately remove from the piles. They may either take as many matches as they like from one pile or take the same number from both piles. The winner is the one who takes the last match.

Determine the initial conditions for which the first player can force a win, and those for which he cannot, and give the correct method of play for him in the former case.

IX. POLYNOMIALS WITH MINIMUM DEVIATION FROM ZERO (TCHEBYCHEV POLYNOMIALS)

Problems 133 to 138 are concerned with P. L. Tchebychev's classical theorem on the polynomials with minimum deviation from zero (see problem 135) and related results. These questions play an important part in modern mathematics.

[6] Nim and Wythoff's games are both played in China, the latter under the name of Tsan-shitsi (which means "choosing stones"). The mathematical theory of the game was given by Wythoff in Ref. [25]. See also Coxeter, Ref. [5].

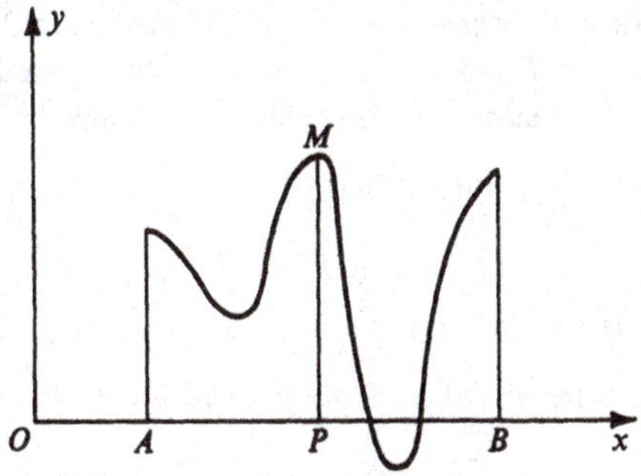

Fig. 16

The *deviation* of a function $f(x)$ from zero on some interval is the maximum absolute value assumed by the function on the interval. Thus the deviation from zero of the function $y = f(x)$, whose graph is given in fig. 16, is the length MP.

132. Show that

a. $$\sin n\alpha = \binom{n}{1} \sin\alpha \cos^{n-1}\alpha - \binom{n}{3} \sin^3\alpha \cos^{n-3}\alpha + \binom{n}{5} \sin^5\alpha \cos^{n-5}\alpha - \cdots;$$

so that, for example,

$$\sin 6\alpha = 6 \sin\alpha \cos^5\alpha - 20 \sin^3\alpha \cos^3\alpha + 6 \sin^5\alpha \cos\alpha.$$

b. $$\cos n\alpha = \cos^n\alpha - \binom{n}{2} \sin^2\alpha \cos^{n-2}\alpha + \binom{n}{4} \sin^4\alpha \cos^{n-4}\alpha - \cdots;$$

so that, for example,

$$\cos 6\alpha = \cos^6\alpha - 15 \sin^2\alpha \cos^4\alpha + 15 \sin^4\alpha \cos^2\alpha - \sin^6\alpha.$$

c. $$\tan n\alpha = \frac{\binom{n}{1} \tan\alpha - \binom{n}{3} \tan^3\alpha + \binom{n}{5} \tan^5\alpha - \cdots}{1 - \binom{n}{2} \tan^2\alpha + \binom{n}{4} \tan^4\alpha - \cdots};$$

so that, for example,

$$\tan 6\alpha = \frac{6 \tan\alpha - 20 \tan^3\alpha + 6 \tan^5\alpha}{1 - 15 \tan^2\alpha + 15 \tan^4\alpha - \tan^6\alpha}.$$

133. *Tchebychev's polynomials.* Prove that for all x in the interval $-1 \leq x \leq 1$, the expression[7]

$$T_n(x) = \cos(n \cos^{-1} x)$$

[7] Here $\cos^{-1} x$ means the angle α such that $0 \leq \alpha \leq \pi$ and $\cos\alpha = x$.

is a polynomial in x of degree n with leading coefficient 2^{n-1}. Find all the roots of the equation $T_n(x) = 0$ and all the values of x between -1 and $+1$ for which $T_n(x)$ assumes a maximum or a minimum.

134. Find the quadratic polynomial

$$x^2 + px + q$$

whose deviation from zero on the interval $-1 \leqq x \leqq 1$ is least.

135.** Prove that the deviation from zero of the polynomial

$$x^n + a_{n-1}x^{n-1} + a_{n-2}x^{n-2} + \cdots + a_1 x + a_0$$

of degree n with leading coefficient 1, on the interval $-1 \leqq x \leqq 1$, is at least $1/2^{n-1}$ and is equal to $1/2^{n-1}$ if and only if the polynomial is $(1/2^{n-1})T_n(x)$ (see problem 133).

136.* Find all the monic polynomials (i.e. with leading coefficient 1) whose deviation from zero on the interval $-2 \leqq x \leqq 2$ is the least possible.

137.** The deviation from zero of a function $f(x)$ on a set of points $x = a, x = b, x = c, \ldots, x = k$ is the maximum of the numbers $|f(a)|$, $|f(b)|, \ldots, |f(k)|$.

Find the monic polynomial of degree n whose deviation from zero on the n points $x = 0, 1, 2, \ldots, n$ is the least possible.

138.*** Let $A_1, A_2, \ldots, A_n$ be any n points of the plane. Show that on any segment of length l there is a point M such that

$$\overline{MA_1} \cdot \overline{MA_2} \cdots \overline{MA_n} \geqq 2\left(\frac{l}{4}\right)^n.$$

How should the n points $A_1, A_2, \ldots, A_n$ be chosen so that on a given segment PQ of length l there is no point M for which

$$\overline{MA_1} \cdot \overline{MA_2} \cdots \overline{MA_n} > 2\left(\frac{l}{4}\right)^n \quad ?$$

X. FOUR FORMULAS FOR π

It has been proved that π, the ratio of the circumference of a circle to its diameter, is not only irrational but is not even a root of any polynomial $x^n + a_1x^{n-1} + \cdots + a_n = 0$ with rational coefficients $a_1, \ldots, a_n$. In particular, there is no expression for π involving only a finite number of rational numbers, addition, subtraction, multiplication, division, and root extraction signs. There are, however, a large number of expressions for π involving, *infinite* sums or products, the first such formula having been obtained in the sixteenth century (see problem 144**a**). In this section we derive a number of the classical formulas for π, allowing us to calculate its value as closely as we like. See Ref. [11].

139. Prove that

a. If $0 < \alpha < \pi/2$, then $\sin \alpha < \alpha < \tan \alpha$.

b. If n is an integer >1, and $0 < n\alpha < \pi/2$, then $(\sin \alpha)/\alpha > (\sin n\alpha)/n\alpha$.

140. Simplify the expression

$$\cos\frac{\alpha}{2}\cos\frac{\alpha}{4}\cos\frac{\alpha}{8}\cdots\cos\frac{\alpha}{2^n}.$$

141. Find polynomials with rational coefficients whose roots are

a. $\cot^2\frac{\pi}{2m+1},\quad \cot^2\frac{2\pi}{2m+1},\quad \cot^2\frac{3\pi}{2m+1},\ldots,\cot^2\frac{m\pi}{2m+1};$

b. $\cot\frac{\pi}{4n},\quad -\cot\frac{3\pi}{4n},\quad \cot\frac{5\pi}{4n},\quad -\cot\frac{7\pi}{4n},\ldots,\cot\frac{(2n-3)\pi}{4n},$

$$-\cot\frac{(2n-1)\pi}{4n}\qquad \text{(for } n \text{ even);}$$

c. $\sin^2\frac{\pi}{2m},\quad \sin^2\frac{2\pi}{2m},\quad \sin^2\frac{3\pi}{2m},\ldots,\sin^2\frac{(m-1)\pi}{2m};$

d. $\sin^2\frac{\pi}{4m},\quad \sin^2\frac{3\pi}{4m},\quad \sin^2\frac{5\pi}{4m},\ldots,\sin^2\frac{(2m-1)\pi}{4m}.$

142. Prove that

a.
$$\cot^2\frac{\pi}{2m+1}+\cot^2\frac{2\pi}{2m+1}+\cot^2\frac{3\pi}{2m+1}+\cdots+\cot^2\frac{m\pi}{2m+1}=\frac{m(2m-1)}{3};$$

b.
$$\csc^2\frac{\pi}{2m+1}+\csc^2\frac{2\pi}{2m+1}+\csc^2\frac{3\pi}{2m+1}+\cdots+\csc^2\frac{m\pi}{2m+1}=\frac{m(2m+2)}{3};$$

c. for even n

$$\cot\frac{\pi}{4n}-\cot\frac{3\pi}{4n}+\cot\frac{5\pi}{4n}-\cot\frac{7\pi}{4n}+\cdots+\cot\frac{(2n-3)\pi}{4n}-\cot\frac{(2n-1)\pi}{4n}=n.$$

143. Prove that

$$\sin\frac{\pi}{2m}\sin\frac{2\pi}{2m}\sin\frac{3\pi}{2m}\cdots\sin\frac{(m-1)\pi}{2m}=\frac{\sqrt{m}}{2^{m-1}}$$

and

$$\sin\frac{\pi}{4m}\sin\frac{3\pi}{4m}\sin\frac{5\pi}{4m}\cdots\sin\frac{(2m-1)\pi}{4m}=\frac{\sqrt{2}}{2^{m}}.$$

144a. From the result of problem 140 deduce *Vieta's*[8] formula

$$\frac{\pi}{2}=\frac{1}{\sqrt{\frac{1}{2}}\sqrt{\frac{1}{2}+\frac{1}{2}\sqrt{\frac{1}{2}}}\sqrt{\frac{1}{2}+\frac{1}{2}\sqrt{\frac{1}{2}+\frac{1}{2}\sqrt{\frac{1}{2}}}}\cdots}.$$

b. Evaluate the infinite product

$$\tfrac{1}{2}\cdot\sqrt{\tfrac{1}{2}+\tfrac{1}{2}\cdot\tfrac{1}{2}}\sqrt{\tfrac{1}{2}+\tfrac{1}{2}\sqrt{\tfrac{1}{2}+\tfrac{1}{2}\cdot\tfrac{1}{2}}}\sqrt{\tfrac{1}{2}+\tfrac{1}{2}\sqrt{\tfrac{1}{2}+\tfrac{1}{2}\sqrt{\tfrac{1}{2}+\tfrac{1}{2}\cdot\tfrac{1}{2}}}}\cdots.$$

145a. From the identities 142**a** and **b** deduce *Euler's* formula

$$\frac{\pi^2}{6}=1+\frac{1}{2^2}+\frac{1}{3^2}+\frac{1}{4^2}+\cdots.$$

b. What is the sum of the infinite series

$$1+\frac{1}{2^4}+\frac{1}{3^4}+\frac{1}{4^4}+\cdots?$$

146a. From the identity of 142c deduce *Leibniz's*[9] formula

$$\frac{\pi}{4}=1-\frac{1}{3}+\frac{1}{5}-\frac{1}{7}+\cdots.$$

b. What is the sum of the infinite series

$$1+\frac{1}{3^2}+\frac{1}{5^2}+\frac{1}{7^2}+\cdots?$$

147. From the identity in problem 143 deduce Wallis's formula[10]

$$\frac{\pi}{2}=\frac{2}{1}\cdot\frac{2}{3}\cdot\frac{4}{3}\cdot\frac{4}{5}\cdot\frac{6}{5}\cdot\frac{6}{7}\cdots.$$

[8] François Vieta (1540–1603), a French mathematician, one of the creators of modern algebraic notation.

[9] Gottfried Wilhelm Leibniz (1646–1716), a German mathematician, one of the inventors of the differential and integral calculus.

[10] John Wallis (1616–1703), an English mathematician.

XI. THE CALCULATION OF AREAS OF REGIONS BOUNDED BY CURVES

Problems 148 to 157 are concerned with the calculation of the areas of curvilinear figures (that is, figures bounded by curves). In high-school geometry we learn to calculate the areas of certain simple plane figures: circles, and sectors and segments of circles. But in many problems in both pure and applied mathematics, we have to calculate the areas of more complicated figures, with boundaries that need not consist of straight lines and arcs of circles. The integral calculus is concerned in part with general

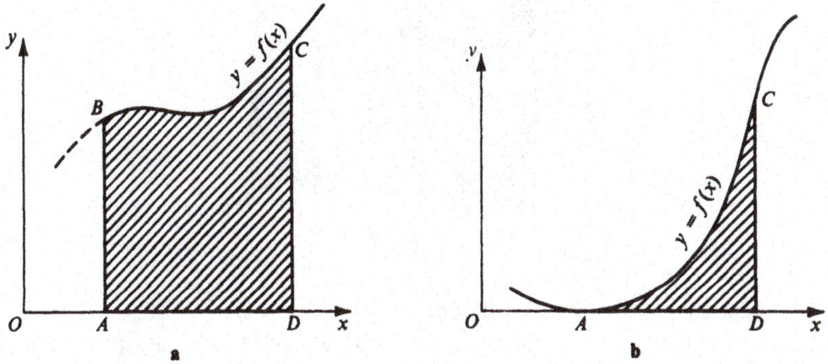

Fig. 17

methods for calculating such areas. Most of these general methods were created in the seventeenth and eighteenth centuries, when the development of the sciences made numerous calculations of this type necessary. However, isolated problems in the calculation of areas were dealt with long before that time: a number of them were solved by various special tricks which use no more than high-school mathematics. We give a number of these below.

The central role in this series of questions is played by problems 151 to 154, which contain the geometric theory of natural logarithms.

In the following problems we will calculate the areas of certain curvilinear trapezoids $ABCD$ bounded by a curve BC given by the equation $y = f(x)$ (for instance, the parabola $y = x^2$, or the sine curve $y = \sin x$), a segment AD of the x axis and two vertical lines AB and CD, corresponding to $x = a$ and $x = b$ (fig. 17a). In certain cases the side AB of our trapezoid may degenerate to a point, and we shall then be dealing with a curvilinear triangle instead of a curvilinear trapezoid. See, for example, triangle ACD in fig. 17b.

In calculating the area of a curvilinear trapezoid $ABCD$ (as in the calculation of the area of a circle) we must use the concept of limit. Let us divide the base line AD of the trapezoid into n parts by means of points $M_1, M_2, \ldots, M_{n-1}$, and let the lengths of $AM_1, M_1M_2, M_2M_3, \ldots, M_{n-1}D$ be $h_1, h_2, h_3, \ldots, h_n$, respectively. Next draw vertical lines through $M_1, M_2, \ldots, M_{n-1}$ to meet the curve $y = f(x)$ at the points $N_1, N_2, \ldots, N_{n-1}$ (fig. 18). Construct n rectangles with bases $AM_1, M_1M_2, M_2M_3, \ldots, M_{n-1}D$ and heights $M_1N_1, M_2M_2, M_3M_3, \ldots, DC$, respectively.

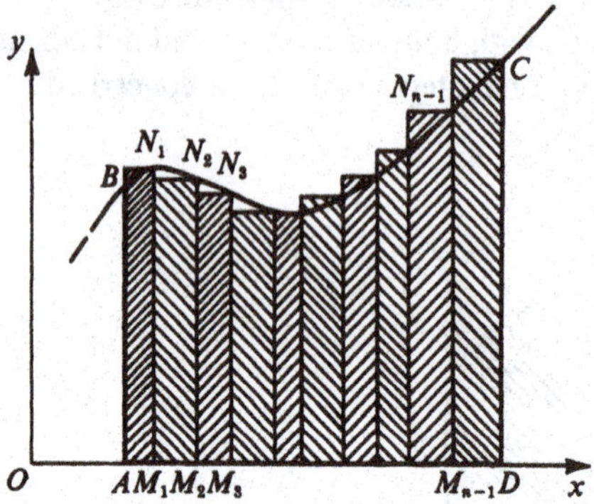

Fig. 18

Let the x coordinates of the points $M_1, M_2, \ldots, M_{n-1}, D$ be $x_1, x_2, \ldots, x_{n-1}, x_n$; then since the x coordinate of A is a, and that of B is b, we have $x_1 = a + h_1, x_2 = a + h_1 + h_2, \ldots, x_{n-1} = a + h_1 + h_2 + \cdots + h_{n-1} = b - h_n,\ x_n = b$. The lengths of the segments $M_1N_1, M_2N_2, M_3N_3, \ldots, DC$ are equal to $f(x_1), f(x_2), f(x_3), \ldots, f(x_n)$, respectively; it follows that the area of the steplike polygon we have constructed from our n rectangles is

$$S_n = f(x_1)h_1 + f(x_2)h_2 + f(x_3)h_3 + \cdots + f(x_n)h_n. \tag{7}$$

New let n tend to infinity, at the same time making all the lengths $h_1, h_2, h_3, \ldots, h_n$ of the segments $AM_1, M_1M_2, M_2M_3, \ldots, M_{n-1}D$ tend to zero. Then in all cases considered in this book, S_n will tend to a limit which is the area of the curvilinear trapezoid $ABCD$.[11]

[11] We might equally well have taken the heights of our n rectangles to be the segments $AB, M_1N_1, M_2N_2, \ldots, M_{n-1}N_{n-1}$. It turns out in all the examples given below that the area of the polygon formed in this way will tend to the same limit as S_n (for $n \to \infty$ and all h_i tending to zero). This common limit is the area of the curvilinear trapezoid $ABCD$.

It follows that we need only distribute our $n - 1$ points $M_1, M_2, \ldots, M_{n-1}$ on AD in a manner that allows us to calculate the sum (7); we then let $n \to \infty$ in the formula

In the following problems, a and b are positive real numbers such that $a < b$.

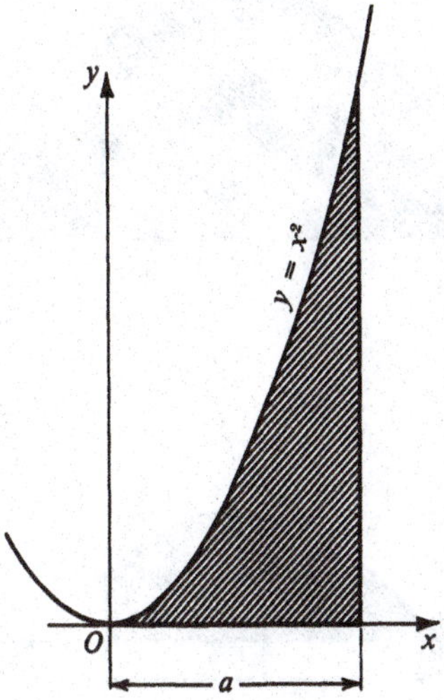

Fig. 19

148. Calculate the area of the curvilinear triangle bounded by the parabola $y = x^2$, the x axis, and the line $x = a$ (fig. 19).

149a. Find the area bounded by the wave ABC of the sine curve $y = \sin x$ $(-\pi/2 \leqq x \leqq 3\pi/2)$ and the line AC (fig. 20a).

b. Find the area of the curvilinear triangle bounded by the sine curve $y = \sin x$, the x axis, and the line $x = a$, where $a < \pi$ (fig. 20b).

150a.* Find the area of the curvilinear trapezoid bounded by the curve $y = x^m$ (where m is an integer $\neq -1$), the x axis, and the lines $x = a$ and $x = b$.

b. Using the result of part **a**, find the area of the curvilinear triangle bounded by the curve $y = x^m$ $(m > 0)$, the x axis, and the line $x = b$.

The next four problems deal with the calculation of the area of the curvilinear trapezoid bounded by the curve $y = x^{-1} = 1/x$, the x axis, and the lines $x = a$ and $x = b$, that is, with the case $m = -1$ excluded in 150**a**.

Logarithms and exponentials are often introduced in high school in the following way. If n is a positive integer, one defines

$$a^n = \overbrace{a\,a\cdots a}^{n \text{ factors}}.$$

---

obtained, and find the area of the curvilinear figure. Of course we must be careful to choose the division points in such a way that as $n \to \infty$ the distance between neighboring points tends to zero.

The calculation of the area of a curvilinear triangle (fig. 17) is treated in the same manner as the calculation of the area of a curvilinear trapezoid.

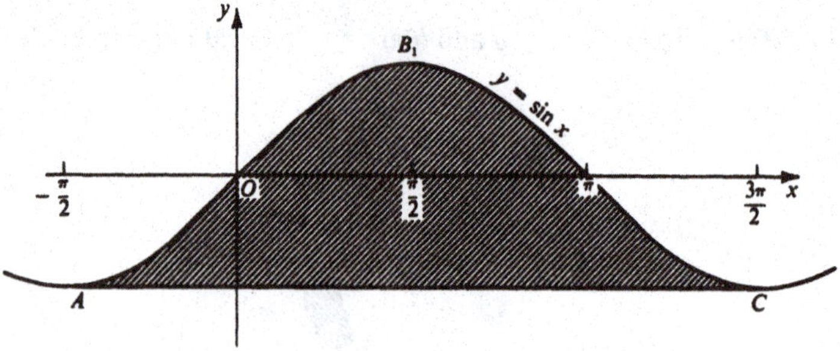

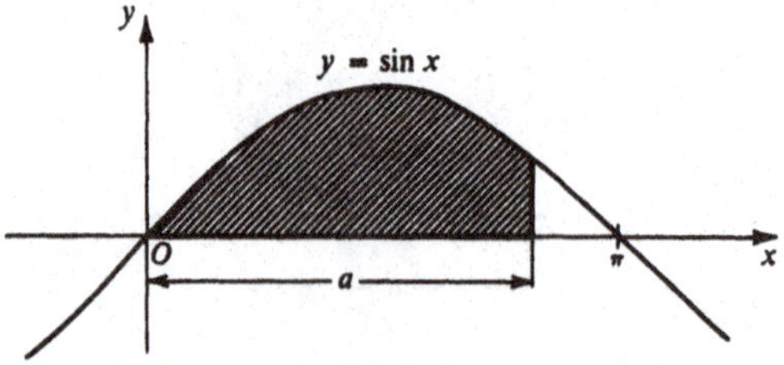

Fig. 20

If $a \neq 0$, and n is a negative integer, then a^n is defined to be $1/a^{-n}$, while a^0 is defined to be 1. It is then shown that $a^{m+n} = a^m \cdot a^n$ and $a^{mn} = (a^m)^n$ for any integers m and n. At this point the student is expected to see intuitively that if $a > 0$, then a^x can be defined even when x is not an integer, in such a way that the "laws of exponents" $a^{u+v} = a^u a^v$ and $a^{uv} = (a^u)^v$ continue to hold. The logarithm of y to the base a, denoted by $\log_a y$ is then defined to be "that number x such that $a^x = y$." The crucial point, namely the existence and uniqueness of such a number x, is not discussed. One then goes on to show that

$$\log_a uv = \log_a u + \log_a v$$

and

$$\log_a u^v = v \log_a u.$$

All of this can be made perfectly rigorous; we will briefly indicate some of the steps. First of all, if q is a positive integer, the function $y = x^q$ increases continuously from 0 to ∞ as x goes from 0 to ∞ (fig. 21).

Hence any horizontal line $y = a$, where $a > 0$, cuts the curve in a unique point. In other words there is a unique value of x such that $x^q = a$. We use the symbol $a^{1/q}$ to denote this value. Next, if p is any integer, and q is a positive integer, we define $a^{p/q}$ to be $(a^{1/q})^p$. Thus we have defined a^r for all rational numbers r. The laws of exponents, namely $a^{r+s} = a^r \cdot a^s$, $a^{rs} = (a^r)^s$, where r and s are rational, are not hard to prove using the above definitions. It is also easy to show that a^r is an increasing

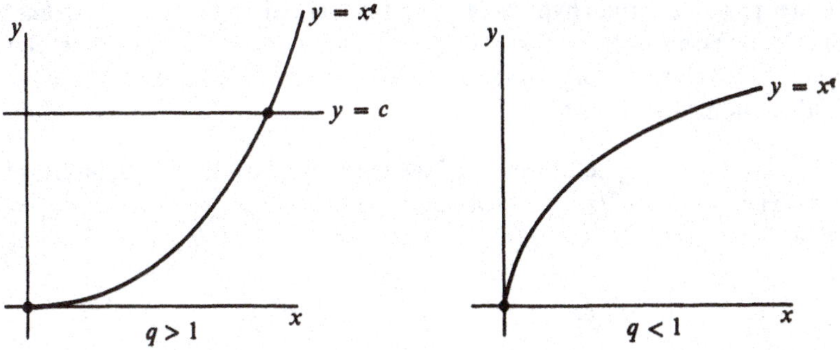

a. b.

Fig. 21

function of r if $a > 1$, i.e. that $a^r < a^s$ whenever $r < s$. If $a < 1$, then a^r is a decreasing function of r.

If x is irrational, choose an increasing sequence $\{r_n\}$ of rational numbers such that $\lim_{n\to\infty} r_n = x$. The sequence $\{a^{r_n}\}$ is monotone and bounded. Hence $\lim_{n\to\infty} a^{r_n}$ exists. We denote this limit by a^x. It can then be shown that the function a^x is continuous and satisfies the laws of exponents. It is monotone increasing if $a > 1$, and decreasing if $a < 1$ (fig. 22).

The continuity and monotonicity of a^x imply that for any given $y > 0$, there is exactly one value of x such that $y = a^x$. We denote this x by the symbol $\log_a y$. The properties $\log_a uv = \log_a u + \log_a v$, $\log_a u_2 = v \log_a u$ are now simple consequences of the laws of exponents.

The development of exponentials and logarithms sketched above can be regarded as the "old-fashioned" treatment. At the present time the preferred method is to define the logarithmic function $\log_a x$ as the area

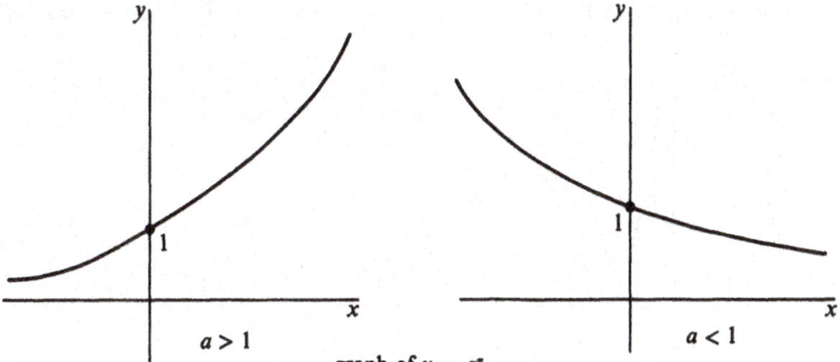

graph of $y = a^x$

a. b.

Fig. 22

under part of a certain hyperbola. The exponential function is then defined as the inverse function of $y = \log_a x$. In problems 151—154 we develop this approach more fully, and show that it leads to the same results as the old-fashioned treatment.

151. Let S_1 and S_2 be the areas of the curvilinear trapezoids bounded by the hyperbola $y = 1/x$, the x axis, and the lines $x = a_1$, $x = b_1$ and $x = a_2$, $x = b_2$, respectively (fig. 23). Show that if $b_1/a_1 = b_2/a_2$, then $S_1 = S_2$.

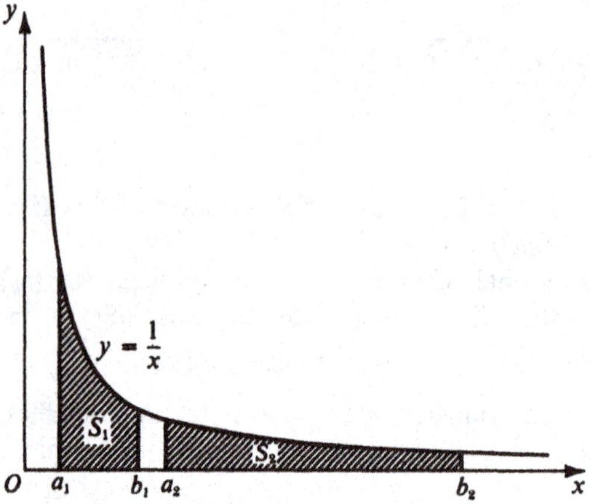

Fig. 23

We now proceed to determine the area of the curvilinear trapezoid bounded by the hyperbola $y = 1/x$, the x axis, and the lines $x = a$ and $x = b$ (fig. 24). By the result of 151 this area depends only on the ratio $b/a = c$; trapezoids for which this ratio is the same have the same area. In other words, the area is a function of $c = b/a$; we denote the function by $F(c)$. It is clear that for every $z > 1$, $F(z)$ is equal to the area of the curvilinear trapezoid bounded by the hyperbola $y = 1/x$, the x axis, and

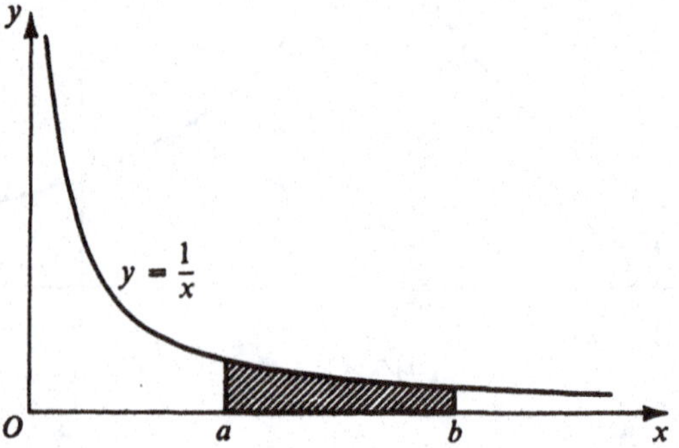

Fig. 24

the lines $x = 1$ and $x = z$ (see fig. 25a, in which this area is shaded). It is natural to put $F(1) = 0$; and in the future we shall do so. As regards values of z less than 1, we agree that $F(z)$ will then mean the area of the trapezoid bounded by the hyperbola, the x axis, and the lines $x = 1$ and $x = z$, but *taken with the negative sign.* We have now defined $F(z)$ for all positive z, and according to our definition $F(z) > 0$ for $z > 1$, $F(1) = 0$, and $F(z) < 0$ for $z < 1$.

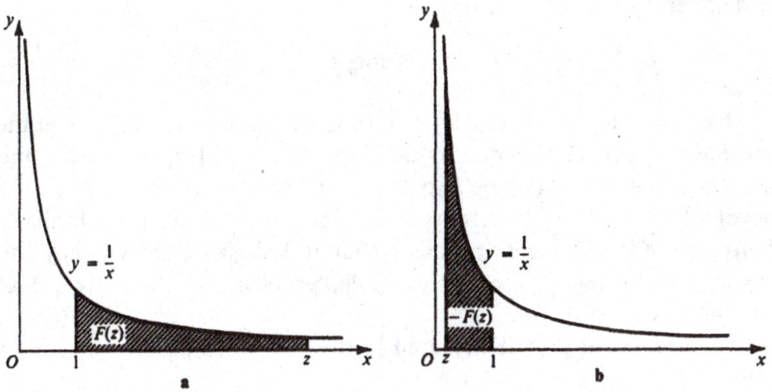

Fig. 25

The following three problems are devoted to a study of the function $F(z)$: they lead us to the conclusion that this function coincides with the logarithmic function, familiar from high-school work, the only difference being that it is to a base different from 10.

152. Prove that for any positive z_1 and z_2

$$F(z_1 z_2) = F(z_1) + F(z_2).$$

153. Prove that the function $F(z)$ assumes the value 1 at some point between 2 and 3.

In what follows we shall always use the letter e to denote the value of z for which $F(z) = 1$. Thus the conclusion of 153 is that the number e exists and that $2 < e < 3$.

This number e plays an important part in mathematics and often appears in contexts that at first glance have nothing to do with its definition as the area under a hyperbola. See, for example, problems 158, 163, 164, and 80 (in vol. I).

154. Suppose $\log_e z$ is defined as indicated on pp. 27–29. Show that

$$F(z) = \log_e z.$$

Thus geometric considerations of the area under a hyperbola have led us to the function[12] $\log z$. This fact will help to explain why the creators of the theory of logarithms, Napier and Bürgi, who developed the theory independently and almost at the same time, both chose as their base not 10 (which might seem the logical choice) but the irrational number e. Napier and Bürgi did not, in fact, consider the area under a hyperbola, but in essence their definitions of logarithms are fairly close to ours, and both immediately lead to logarithms to base e.[13]

Logarithms to base e are customarily called *natural logarithms* and are denoted by the symbol ln:

$$\ln z = \log_e z.$$

We have thus proved that $F(z) = \ln z$. It follows that the area of the curvilinear trapezoid bounded by the hyperbola $y = 1/x$, the x axis, and the lines $x = a$, $x = b$ (where $b > a$) is equal to $\ln (b/a)$ (see problem 151 above). This result has numerous applications, and it explains the frequent occurrence of logarithms in problems that at first glance seem to have no connection with them. See, for example, problems 155, 156, 167, 170, 173, and 174.

For notions of probability used in the next three problems, refer to Section VII (page 27) in Volume I.

155.*** A rod is broken into three pieces; the two break points are chosen at random. What is the probability that an acute-angled triangle can be formed from the three pieces?

156.*** A rod is broken in two at a point chosen at random; then the larger of the two pieces is broken in two at a point chosen at random. What is the probability that the three pieces obtained can be joined to form a triangle?

157.*** *Buffon's problem.*[14] A plane is ruled with parallel lines, the distance between two consecutive lines being $2a$. A needle of length $2a$ and negligible thickness is dropped in a random fashion on the plane surface. Prove that the probability that the needle lands across one of the lines is $2/\pi \approx 0.637$ (where π as usual denotes the ratio of the circumference of a circle to its diameter).

[12] We can arrive geometrically at logarithms to other bases simply by considering the area under hyperbolas $y = c/x$, where $c \neq 1$. For instance, we arrive at the so-called common logarithms by choosing $c = 1/\log_e 10 \approx 0.4343$. However, there are a number of reasons why logarithms to base e are particularly simple and natural. These reasons are connected with the fact that the simplest hyperbola of the form $y = c/x$ is the one with $c = 1$.

[13] Reference [1] deals with the history of the theory of logarithms.

[14] Georges Buffon (1707–1788), a famous French scientist.

XII. SOME REMARKABLE LIMITS

The following problems on limits are connected with the geometric problems of the last section. The solutions depend on the results of problems 148 to 154. Purely algebraic solutions to some of these problems are given in references [17] and [21].

158. Find the values of the following limits:

a. $\lim_{n\to\infty} n \ln\left(1 + \frac{1}{n}\right)$.

b. $\lim_{n\to\infty} n \log_a\left(1 + \frac{1}{n}\right)$.

c. $\lim_{n\to\infty} n(\sqrt[n]{a} - 1)$.

159a.* Find the area of the curvilinear trapezoid bounded by the curve $y = a^x$, the x axis, the y axis, and the line $x = b$ (fig. 26a).

b. Find the area of the curvilinear triangle bounded by the curve $y = \log_a x$, the x axis, and the line $x = b$, where $b > 1$ (fig. 26b).

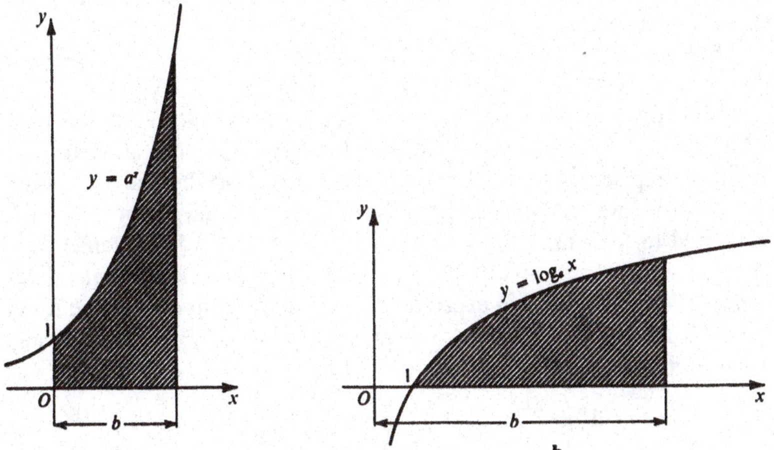

Fig. 26

160.* Find the area of the curvilinear trapezoid bounded by the curve $y = (\log_a x)/x$, the x axis, and the line $x = b$, where $b > 1$ (fig. 27).

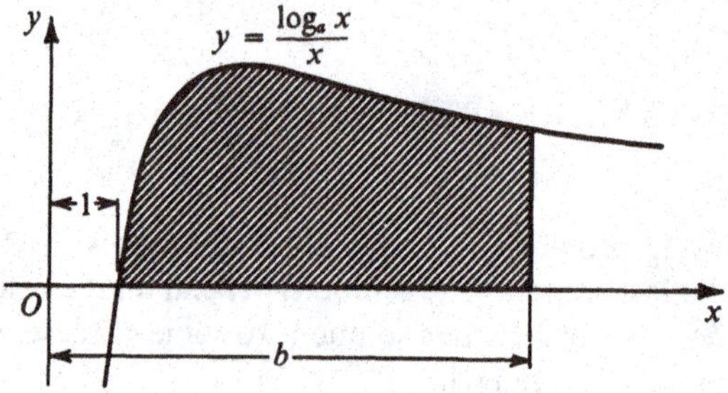

Fig. 27

161. Show that for $k > -1$

$$\lim_{n\to\infty}\frac{1^k + 2^k + 3^k + \cdots + n^k}{n^{k+1}} = \frac{1}{k+1}.$$

The result of 161 shows that for large n the sum $1^k + 2^k + 3^k + \cdots + n^k$, where $k > -1$, is asymptotically $n^{k+1}/(k + 1)$. This same sum for the cases $k = -1$ and $k < -1$ will be considered in problems 167 and 169.

In what follows we repeatedly come across approximate estimates for sums and products, made on the assumption that the number of summands or factors is large. These estimates are of two different types. If we are lucky, we find a comparatively simple expression such that the difference between it and the sum of n terms tends to zero as $n \to \infty$. In such cases the *absolute error* made by replacing the sum by our expression will be very small for large n. We shall say that the sum of n terms is *approximately equal* to our expression and shall use the symbol $\approx$ for approximate equality.

But sometimes, for example in problem 161, we have a different situation. Here we cannot assert that the difference $1^k + 2^k + 3^k + \cdots + n^k - n^{k+1}/(k + 1)$ becomes very small for large n. However, the *ratio* $(1^k + 2^k + 3^k + \cdots + n^k)(k + 1)/n^{k+1}$ for large n is very close to one. So if we replace the sum $1^k + 2^k + 3^k + \cdots + n^k$ by the expression $n^{k+1}/(k + 1)$, we commit an error which may be large, but is small compared to the sum $1^k + 2^k + 3^k + \cdots + n^k$ itself. The *relative* error approaches zero as $n \to \infty$. In such cases mathematicians speak of the asymptotic equality of two expressions and use the symbol $\sim$. The result of problem 161 is that the sum $1^k + 2^k + 3^k + \cdots + n^k$ is asymptotically equal to $n^{k+1}/(k + 1)$:

$$1^k + 2^k + 3^k + \cdots + n^k \sim \frac{1}{k+1}n^{k+1}.$$

162a.* Show that the sequence

$$\left(1+\frac{1}{1}\right)^1,\ \left(1+\frac{1}{2}\right)^2,\ \left(1+\frac{1}{3}\right)^3, \ldots, \left(1+\frac{1}{n}\right)^n, \ldots$$

is increasing.

b. Show that the sequence

$$\left(1+\frac{1}{1}\right)^2, \left(1+\frac{1}{2}\right)^3, \left(1+\frac{1}{3}\right)^4, \ldots, \left(1+\frac{1}{n}\right)^{n+1}, \ldots$$

is decreasing.

From the results of 159**a** and **b** it follows easily that the two sequences considered in them both tend to one and the same limit. For by **a** the sequence

$$\left(1+\frac{1}{1}\right)^1, \left(1+\frac{1}{2}\right)^2, \left(1+\frac{1}{3}\right)^3, \ldots$$

is increasing. Moreover, the terms cannot increase indefinitely, since they are all less than $(1 + 1/1)^2 = 4$. This follows from the inequality

$$\left(1+\frac{1}{n}\right)^n < \left(1+\frac{1}{n}\right)^{n+1} \leqq \left(1+\frac{1}{1}\right)^2,$$

which in turn follows from the fact that the sequence in **b** is decreasing. Thus the terms of the first sequence tend to a limit. Similarly, we can show that the terms of the second sequence tend to a limit: its terms are decreasing, and each is greater than $(1 + 1/1)^1 = 2$, since

$$\left(1+\frac{1}{n}\right)^{n+1} > \left(1+\frac{1}{n}\right)^n \geqq \left(1+\frac{1}{1}\right)^1.$$

Moreover, since

$$\lim_{n\to\infty} \frac{(1 + 1/n)^{n+1}}{(1 + 1/n)^n} = \lim_{n\to\infty} \left(1+\frac{1}{n}\right) = 1,$$

the limits of the two sequences coincide.

This common limit lies between 2 and 4; in fact, it is just the number e, defined geometrically in problem 153:

$$\lim_{n\to\infty} \left(1+\frac{1}{n}\right)^n = \lim_{n\to\infty} \left(1+\frac{1}{n}\right)^{n+1} = e.$$

(See the solution of problem 163.)

From the results of **a** and **b** it follows that for all positive integers n

$$\left(1+\frac{1}{n}\right)^n < e < \left(1+\frac{1}{n}\right)^{n+1};$$

this allows us to determine the value of e to any desired degree of accuracy. The decimal expansion of e starts $e = 2.718281828459045\ldots$. In practice one would not use the result of problem 162 to calculate e, but the infinite series given below in problem 164.

163.* Show that for any positive or negative z

$$e^z = \lim_{n\to\infty}\left(1+\frac{z}{n}\right)^n$$

164.** Show that

$$e^z = 1 + z + \frac{z^2}{2!} + \frac{z^3}{3!} + \cdots + \frac{z^n}{n!} + \cdots$$

so that in particular (for $z = 1$ and $z = -1$),

$$e = 1 + 1 + \frac{1}{2!} + \frac{1}{3!} + \cdots + \frac{1}{n!} + \cdots$$

$$\frac{1}{e} = e^{-1} = \frac{1}{2!} - \frac{1}{3!} + \cdots + \frac{(-1)^n}{n!} + \cdots.$$

165.*** Show that for any positive integer n the number $n!$ satisfies the inequality

$$\sqrt{\frac{4}{5}}\,e\cdot\sqrt{n}\left(\frac{n}{e}\right)^n < n! < e\cdot\sqrt{n}\left(\frac{n}{e}\right)^n.$$

166a.*** Show that the ratio

$$n!:\sqrt{n}\left(\frac{n}{e}\right)^n$$

tends to a limit C as $n \to \infty$. Note that by the previous result C lies between $\sqrt{4/5}e$ and e, and hence, between 2.43 and 2.72.

b. Show that the number C of part **a** is equal to $\sqrt{2\pi} \approx 2.50$, where π is the ratio of the circumference of a circle to its diameter.

From the result of 166 it follows that

$$\lim_{n\to\infty}\frac{n!}{\sqrt{2\pi n}(n/e)^n} = 1,$$

in other words, that

$$n! \sim \sqrt{2\pi n}\left(\frac{n}{e}\right)^n.$$

(For the meaning of $\sim$ see the discussion after problem 161.) This approximation formula is known as *Stirling's formula*[15] and is of frequent application in various branches of mathematics and physics. It turns out

[15] James Stirling (1692–1770), a Scottish mathematician.

that for $n = 10$ Stirling's formula already gives a very good approximation to $n!$. (In fact, $10! = 3{,}628{,}800$, and using 5-figure tables we find that $\sqrt{20\pi}\,(10/e)^{10} \approx 3{,}598{,}700$ (to 5 significant figures); thus if we replace $10!$ by $\sqrt{20\pi}\,(10/e)^{10}$ our error is less than 1%.)

As n increases further, the relative error involved in using the formula decreases rapidly. At the same time it is precisely for large values of n that the direct calculation of $n!$ as the product of all the integers from 1 to n becomes laborious. Stirling's formula is used in the remarks following the solutions of problems **78b**, **79a**, and **79b** in the first volume of this book.

167. Put

$$1 + \frac{1}{2} + \frac{1}{3} + \cdots + \frac{1}{n-1} - \ln n = \gamma_n.$$

Show that

a. $0 < \gamma_n < 1$.

b. As $n \to \infty$, γ_n tends to a limit γ.

Thus for large n we have the approximate equality

$$1 + \frac{1}{2} + \frac{1}{3} + \cdots + \frac{1}{n} \approx \ln(n+1) + \gamma, \tag{8}$$

and the accuracy of this approximate equation increases as n increases. As $n \to \infty$ the difference

$$\ln(n+1) - \ln n = \ln\frac{n+1}{n} = \ln\left(1 + \frac{1}{n}\right)$$

tends to zero (since $(1 + 1/n) \to 1$), so that (8) can also be written in the more elegant form

$$1 + \frac{1}{2} + \frac{1}{3} + \cdots + \frac{1}{n} \approx \ln n + \gamma.$$

(Compare this with the result of problem 173.)

The number

$$\gamma = \lim_{n\to\infty}\left(1 + \frac{1}{2} + \frac{1}{3} + \cdots + \frac{1}{n-1} - \ln n\right)$$

appears often in higher mathematics (see, for example, "Mertens' third formula," problem 174); it is called *Euler's constant.* Its decimal expansion starts $\gamma = 0.57721566\ldots$.

168a.* Show that there exists a number C such that the difference

$$\frac{\log 1}{1} + \frac{\log 2}{2} + \frac{\log 3}{3} + \cdots + \frac{\log(n-1)}{n-1} - C\log^2 n = \delta_n$$

always lies between $-\frac{1}{2}$ and $+\frac{1}{2}$. Find this number C.

b. Show that as $n \to \infty$, δ_n tends to a limit δ (which also lies between $-\frac{1}{4}$ and $+\frac{1}{4}$).

Thus for large n we have the approximate equality

$$\frac{\log 1}{1} + \frac{\log 2}{2} + \frac{\log 3}{3} + \cdots + \frac{\log (n-1)}{n-1} \approx C \log^2 n + \delta,$$

or alternatively

$$\frac{\log 1}{1} + \frac{\log 2}{2} + \frac{\log 3}{3} + \cdots + \frac{\log n}{n} \approx C \log^2 n + \delta.$$

The accuracy of the approximation increases with n.

It is interesting to compare this with problem 171.

169. Show that as n tends to infinity, for $s > 1$ the sum

$$1 + \frac{1}{2^s} + \frac{1}{3^s} + \cdots + \frac{1}{n^s}$$

tends to a limit lying between $1/(s-1)$ and $s/(s-1)$.

The proposition of 169 may also be phrased as follows: the *infinite series*

$$1 + \frac{1}{2^s} + \frac{1}{3^s} + \frac{1}{4^s} + \cdots + \frac{1}{n^s} + \cdots \tag{9}$$

converges for $s > 1$, and its sum lies between $1/(s-1)$ and $s/(s-1)$. Thus for $s = 2$ the sum lies between 1 and 2, for $s = 4$ it lies between $\frac{1}{3}$ and $\frac{4}{3}$. For the exact sum of the series $1 + 1/2^2 + 1/3^2 + 1/4^2 + \cdots$ and $1 + 1/2^4 + 1/3^4 + 1/4^4 + \cdots$ see problems 145**a** and **b**. For $s \leq 1$ the infinite series (9) diverges; the partial sums $1 + 1/2^s + \cdots + 1/n^s$ tend to infinity as $n \to \infty$. (See problems 161 and 167.)

XIII. THE THEORY OF PRIMES

The theory of numbers is the branch of mathematics that studies the properties of integers. A large part of the theory is concerned with the study of prime numbers (that is, numbers having no divisors except 1 and themselves); problems 170–174 treat this topic. The central result among these problems is Tchebychev's theorem (problem 170).

Despite the apparent simplicity of the questions that arise in the theory of prime numbers, it remains one of the deepest branches of mathematics. A number of the central questions have been solved only recently, and some are still unsolved. Significant results in the theory of prime numbers have been obtained in the last few decades by I. M. Vinogradov, P. Erdös, and A. Selberg, among others. See Refs. [12], [18], and [23].

The number of primes $\leq N$ is denoted by $\pi(N)$; thus

$\pi(1) = 0$ (1 does not count as a prime), $\pi(2) = 1$, $\pi(3) = \pi(4) = 2$,

$\pi(5) = \pi(6) = 3$, $\pi(7) = \pi(8) = \pi(9) = \pi(10) = 4$, $\pi(11) = \pi(12) = 5$,

$\pi(13) = \pi(14) = \pi(15) = \pi(16) = 6, \ldots$

170.*** *Tchebychev's theorem.* Show that there exist positive constants A and B such that for every N

$$A \frac{N}{\log N} < \pi(N) < B \frac{N}{\log N}.$$

It is clear that the result of problem 93, Volume I, follows from Tchebychev's theorem, since the inequality $\pi(N) < BN/\log N$ implies that $\pi(N)/N \to 0$ as $N \to \infty$.

Tchebychev's theorem asserts that the number $\pi(N)$ of primes $\leq N$ is of the order $N/\log N$. This remarkable theorem was an important step in determining the asymptotic behavior of $\pi(N)$ as $N \to \infty$.

Tchebychev found fairly close limits between which $\pi(N)$ must lie. He showed that

$$0.40 \frac{N}{\log N} < \pi(N) < 0.48 \frac{N}{\log N}.$$

This result may be written more elegantly if we pass over to natural logarithms (logarithms to base $e = 2.718 \ldots$; see problems 151 to 154): $\pi(N)$ satisfies the inequalities

$$0.92 \frac{N}{\ln N} < \pi(N) < 1.11 \frac{N}{\ln N}.$$

Thus we see that $\pi(N)$ is approximated with considerable accuracy by the function $N/\ln N$ (since the numerical multiples 0.92 and 1.11 are both close to 1). Tchebychev also proved that if the ratio $\pi(N) : N/\ln N$ tends to a limit as $N \to \infty$, then this limit is necessarily 1. That the limit of the ratio $\pi(N) : N/\ln N$ as $N \to \infty$ does in fact exist (and therefore is equal to 1) was not proved until the end of the last century, some fifty years after Tchebychev's work. The first proofs of the existence of the limit, due to the French mathematician J. Hadamard and the Belgian mathematician Ch. de la Vallée Poussin, required the application of deep ideas from higher mathematics. An "elementary" (although very complicated) proof was found in 1948 by the Hungarian mathematician P. Erdös and the Norwegian mathematician A. Selberg.

Thus it is now known that

$$\pi(N) \sim \frac{N}{\ln N}$$

(for the meaning of the symbol $\sim$ see above, under problem 161).

171.*** *Mertens' first theorem.*[16] Let 2, 3, 5, 7, 11, . . . , p be the primes not exceeding a given integer N. Show that for all N, the quantity

$$\left|\frac{\log 2}{2} + \frac{\log 3}{3} + \frac{\log 5}{5} + \frac{\log 7}{7} + \frac{\log 11}{11} + \cdots + \frac{\log p}{p} - \log N\right|$$

is bounded, in fact < 4.

As N increases indefinitely, so does its logarithm, for $\log N$ is greater than any given number K as soon as N is greater than 10^K. The sum

$$\frac{\log 2}{2} + \frac{\log 3}{3} + \frac{\log 5}{5} + \cdots + \frac{\log p}{p},$$

where 2, 3, 5, 7, 11, . . . , p are the primes $\leq N$, also tends to infinity with N. Mertens' first theorem asserts that the difference between these two quantities remains bounded: the absolute value of this difference is always less than 4. Thus the sum

$$\frac{\log 2}{2} + \frac{\log 3}{3} + \frac{\log 5}{5} + \frac{\log 7}{7} + \frac{\log 11}{11} + \cdots + \frac{\log p}{p}$$

may be approximated by the simple expression $\log N$. The error is always less than 4, and therefore as $N \to \infty$ the *relative* error tends to 0.

In particular, it follows from Mertens' first theorem that

$$\frac{\log 2}{2} + \frac{\log 3}{3} + \frac{\log 5}{5} + \cdots + \frac{\log p}{p} \sim \log N$$

(for the meaning of the notation see above, in the discussion following problem 161). For as $N \to \infty$

$$\left(\frac{\log 2}{2} + \frac{\log 3}{3} + \frac{\log 5}{5} + \cdots + \frac{\log p}{p}\right)\Big/\log N - 1$$

$$= \left(\frac{\log 2}{2} + \frac{\log 3}{3} + \frac{\log 5}{5} + \cdots + \frac{\log p}{p} - \log N\right)\Big/\log N \to 0.$$

172a. *Abel's formula.*[17] Consider the sum

$$S = a_1b_1 + a_2b_2 + a_3b_3 + \cdots + a_nb_n,$$

[16] F. Mertens, an Austrian mathematician who specialized in the theory of numbers. He was active at the end of the nineteenth century.

[17] Niels Henrik Abel (1802–1829), a brilliant Norwegian mathematician who in the course of his short life established a large number of important results in algebra and analysis.

where $a_1, a_2, a_3, \ldots, a_n$ and $b_1, b_2, b_3, \ldots, b_n$ are any two sequences of numbers. Denote the sums $b_1, b_1 + b_2, b_1 + b_2 + b_3, b_1 + b_2 + \cdots + b_n$ by $B_1, B_2, B_3, \ldots, B_n$, respectively. Show that

$$S = (a_1 - a_2)B_1 + (a_2 - a_3)B_2 + (a_3 - a_4)B_3 + \cdots + (a_{n-1} - a_n)B_{n-1} + a_nB_n.$$

b. Using Abel's formula, calculate the value of

1. $1 + 2q + 3q^2 + \cdots + nq^{n-1}$.
2. $1 + 4q + 9q^2 + \cdots + n^2q^{n-1}$.

173.*** *Mertens' second theorem.* **a.** Let $2, 3, 5, 7, 11, \ldots, p$ be the primes not exceeding the integer N. Show that for all $N > 1$, the expression

$$\frac{1}{2} + \frac{1}{3} + \frac{1}{5} + \frac{1}{7} + \frac{1}{11} + \cdots + \frac{1}{p} - \ln \ln N$$

has absolute value less than some constant T. (We could, for example, take $T = 15$.)

b. Show that the difference

$$\frac{1}{2} + \frac{1}{3} + \frac{1}{5} + \frac{1}{7} + \frac{1}{11} + \cdots + \frac{1}{p} - \ln \ln N$$

tends to a limit β as $N \to \infty$. It follows from part **a** that $\beta \leqq 15$; in fact, β is approximately $\frac{1}{4}$.

We thus have the approximate equality

$$\frac{1}{2} + \frac{1}{3} + \frac{1}{5} + \frac{1}{7} + \frac{1}{11} + \cdots + \frac{1}{p} \approx \ln \ln N + \beta.$$

174. *Mertens' third theorem.* Let $2, 3, 5, 7, 11, \ldots, p$ be the primes not exceeding the integer N. Show that as $N \to \infty$, the product

$$\ln N\left(1 - \frac{1}{2}\right)\left(1 - \frac{1}{3}\right)\left(1 - \frac{1}{5}\right)\left(1 - \frac{1}{7}\right)\left(1 - \frac{1}{11}\right) \cdots \left(1 - \frac{1}{p}\right)$$

tends to a limit c.

We can also write Mertens' third theorem in the form

$$\left(1 - \frac{1}{2}\right)\left(1 - \frac{1}{3}\right)\left(1 - \frac{1}{5}\right)\left(1 - \frac{1}{7}\right)\left(1 - \frac{1}{11}\right) \cdots \left(1 - \frac{1}{p}\right) \sim \frac{c}{\ln N}.$$

Using advanced methods it can be shown that the constant c is equal to $e^{-\gamma}$. Here $e \approx 2.718\ldots$ is the base for the system of natural logarithms, and γ is Euler's constant, defined in problem 167.

We also note the following curious formula, closely connected with Mertens' third theorem:

$$\left(1+\frac{1}{2}\right)\left(1+\frac{1}{3}\right)\left(1+\frac{1}{5}\right)\left(1+\frac{1}{7}\right)\cdots\left(1+\frac{1}{p}\right)\sim\frac{6e^{\gamma}}{\pi^2}\ln N$$

(where 2, 3, 5, 7, . . . , p are the primes not exceeding N). Here e and γ are defined as above, and π is the ratio of the circumference of a circle to its diameter. The preceding formula follows from

$$\left(1+\frac{1}{2}\right)\left(1+\frac{1}{3}\right)\left(1+\frac{1}{5}\right)\left(1+\frac{1}{7}\right)\left(1+\frac{1}{11}\right)\cdots\left(1+\frac{1}{p}\right)$$
$$=\frac{(1-1/2^2)(1-1/3^2)(1-1/5^2)(1-1/7^2)(1-1/11^2)\cdots(1-1/p^2)}{(1-1/2)(1-1/3)(1-1/5)(1-1/7)(1-1/11)\cdots(1-1/p)},$$

and

$$\left(1-\frac{1}{2}\right)\left(1-\frac{1}{3}\right)\left(1-\frac{1}{5}\right)\left(1-\frac{1}{7}\right)\left(1-\frac{1}{11}\right)\cdots\left(1-\frac{1}{p}\right)\sim\frac{e^{-\gamma}}{\ln N}$$

(see problem 174), while

$$\left(1-\frac{1}{2^2}\right)\left(1-\frac{1}{3^2}\right)\left(1-\frac{1}{5^2}\right)\left(1-\frac{1}{7^2}\right)\left(1-\frac{1}{11^2}\right)\cdots\left(1-\frac{1}{p^2}\right)\sim\frac{6}{\pi^2}$$

by Euler's formula, which is given in problem 145a. (See the solution to problem 92 of vol. 1.)

The three theorems of Mertens, together with the prime number theorem (Hadamard-de la Vallée Poussin; see the text following problem 170), are examples of the remarkable connection between natural logarithms and the distribution of the primes among the integers.

SOLUTIONS

SOLUTIONS

I. POINTS AND LINES

101. Yes. Consider, for example, 10 straight lines in the plane, no two of which are parallel and no three of which are concurrent. Let the lines be the bus routes and let the points of intersection be the stops. We can get from any one stop to any other (if the stops lie on one line, without changing; and if not, then with just one change). If we discard one line, it is still possible to get from any one stop to any other, changing buses at most once. However, if we discard two lines, then one stop—their point of intersection—will have no bus routes passing through it, and it will be impossible to get from this stop to any other.

102. One possible arrangement of the routes is shown in fig. 28. There are seven routes and seven stops.

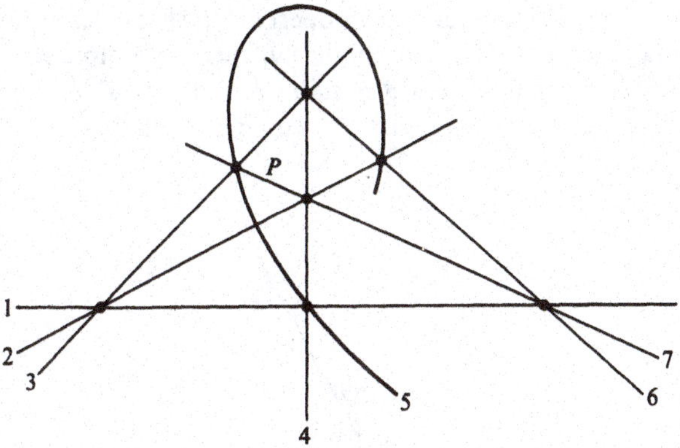

Fig. 28

103a. We note that given any two stops P and Q, there is exactly one line joining them. For by property (2) there is at least one such line. But if there were two distinct lines passing through P and Q, then property (3) would be violated.

Now denote by $f(P)$ the number of lines through the stop P, and by $g(l)$ the number of stops on the line l. We prove first that if P does not lie

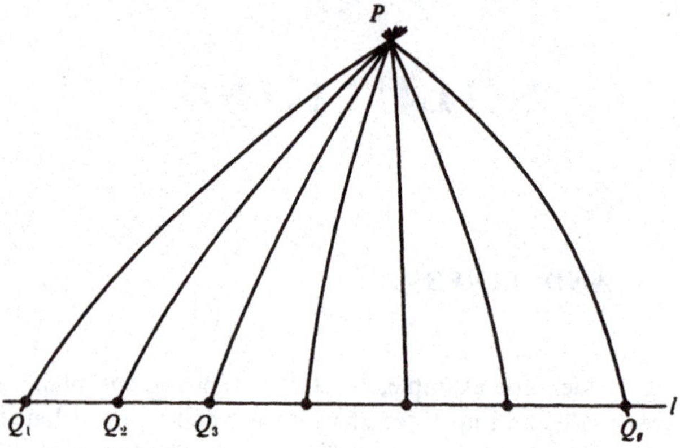

Fig. 29

on l, then $f(P) = g(l)$. For let the stops of l be $Q_1, Q_2, \ldots, Q_g$, where $g = g(l)$. By the preceding remark, there is exactly one line joining P to each of the g stops $Q_1, Q_2, \ldots, Q_g$. But by property (3) *every* line through P has a stop in common with l, that is, every line through P passes through one of the stops $Q_1, Q_2, \ldots, Q_g$. Therefore there are exactly g lines through P, that is, $f(P) = g(l)$.

Now consider any two distinct lines a and b. We will prove that there is a stop P which is on neither a nor b. By property (3), a and b have exactly one stop C in common. By property (1), a has at least three stops, so there is a stop $A \neq C$ on a. Similarly, there is a stop $B \neq C$ on b (fig. 30). The line joining A and B contains a third stop P by property (1). If P were on either a or b, property (3) would be violated.

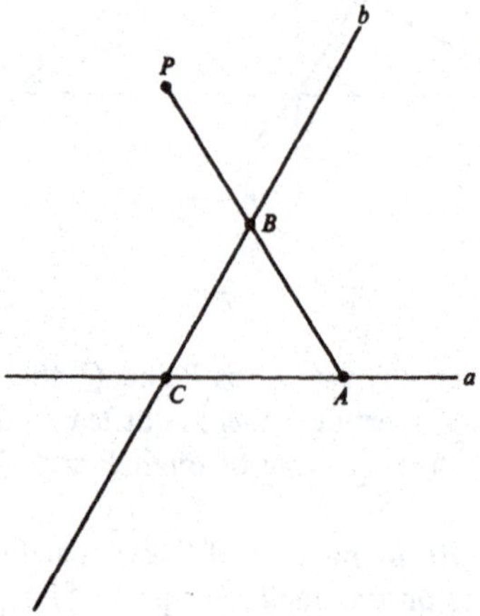

Fig. 30

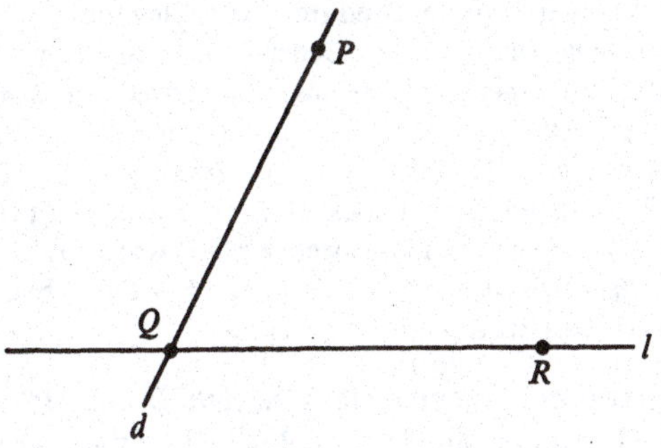

Fig. 31

From what we showed above it now follows that $g(a) = f(P)$ and $g(b) = f(P)$, so that $g(a) = g(b)$. Thus all lines have the same number of stops, which we denote by $n + 1$. (This notation proves to be convenient in more advanced work; see the remark at the end of the problem.)

It remains to be shown that if P is any stop, then $f(P) = n + 1$. It suffices to prove that there is some line l which does not pass through P; we then have $f(P) = g(l) = n + 1$. To construct such a line l, choose a stop $Q \neq P$. Then there is a line d joining P to Q. (fig. 31).

By hypothesis there is more than one line, so there is a stop R not on d. Then the line l joining Q and R has the desired property.

b. Let l be any line, and let $Q_1, Q_2, \ldots, Q_{n+1}$ be the stops on l. Every line other than l has exactly one stop in common with l, that is, goes through exactly one of the stops $Q_1, Q_2, \ldots, Q_{n+1}$. By part **a** we know that through each Q_i there are $n + 1$ lines, one of which is l itself (fig. 32). Hence there are altogether $(n + 1)n$ lines other than l. Together with l, this gives a total of $(n + 1)n + 1 = n^2 + n + 1$ lines.

Similarly, we choose any stop P, and let $l_1, l_2, \ldots, l_{n+1}$ be the lines through P. Since every stop is joined to P by a line, every stop is on one of the lines l_i. By part **a**, l_i has $n + 1$ stops, one of which is P itself. Hence each l_i has n stops other than P, so that there are altogether $(n + 1)n$ stops other than P. Together with P, we get a total of $(n + 1)n + 1 = n^2 + n + 1$ stops.

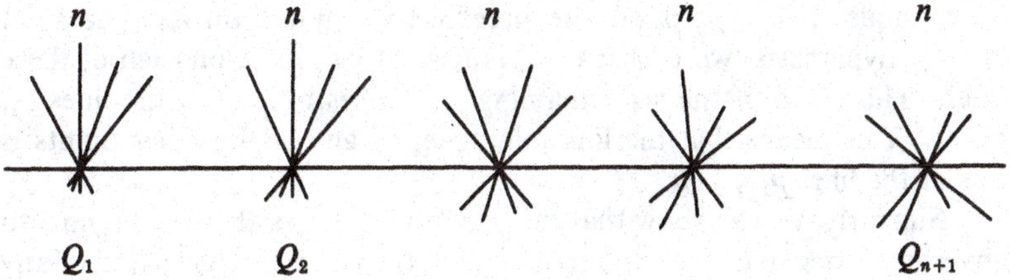

Fig. 32

Remark. A system Π of lines and stops satisfying the conditions of problem 103 is called a *finite projective plane*. Generally the stops are called *points*. The number n is called the *order* of Π. It has been proved that if n is a power of a prime, then there exists a projective plane of order n. Thus there exist planes of orders 2, 3, 4, 5, 7, 8, 9, 11, 13, 16, . . . For example, if $n = 2$, the system of $n^2 + n + 1 = 7$ points and seven lines shown in fig. 28 is a projective plane of order 2. No one has ever found a projective plane whose order n is not a prime power. It has been proved that there is none of order 6, but it is not known whether or not there exists a plane of order 10.

104a. Arrange two congruent rectangles $ABCD$ and $BEFC$ so that they touch along the common side BC (fig. 33). Draw the diagonals of each of these rectangles, and of the large rectangle $AEFD$ which they form. Each of the three pairs of diagonals intersects in a point lying on the straight

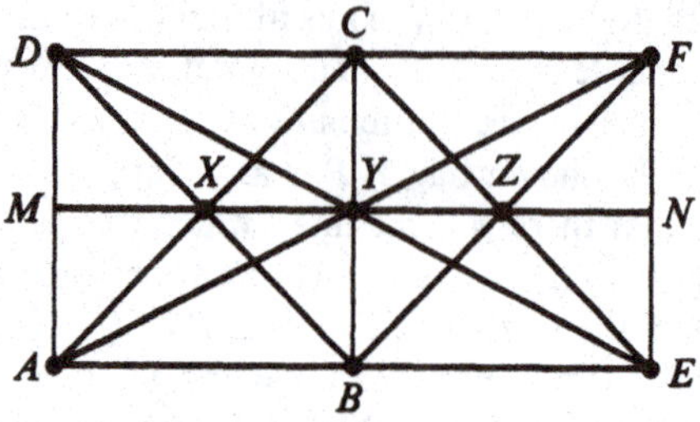

Fig. 33

line MN, where M and N are the midpoints of AD and EF, respectively. These three points of intersection X, Y, Z, together with the points A, B, C, D, E, F, comprise the nine points. As the nine lines we take the six diagonals, the horizontal lines AE and DF, and the line MN, which lies halfway between them. It is easy to see that these nine points and nine lines satisfy all the conditions of the problem.

b. Suppose we could arrange seven points $A_1, A_2, A_3, A_4, A_5, A_6, A_7$ and seven lines $p_1, p_2, p_3, p_4, p_5, p_6, p_7$ in a configuration satisfying the conditions of the problem. We show first that in this case any line joining two of the points $A_1, A_2, A_3, A_4, A_5, A_6, A_7$ is one of the seven lines $p_1, p_2, p_3, p_4, p_5, p_6, p_7$, and that any point of intersection of two of the lines $p_1, p_2, p_3, p_4, p_5, p_6, p_7$ is one of the points $A_1, A_2, A_3, A_4, A_5, A_6, A_7$. Suppose, for example, that p_1, p_2, p_3 are the three lines which pass through the point A_1. By hypothesis two of our points (apart from A_1) lie on each of these lines. Thus each of the six remaining points lies on one of the lines p_1, p_2, p_3. This means that the line joining A_1 to any of the other points is one of the lines p_1, p_2, p_3.

Similarly, we can show that the line joining any of the seven points to any other is one of our seven lines, and that the point of intersection of any two of our seven lines is one of the points $A_1, A_2, A_3, A_4, A_5, A_6, A_7$. For

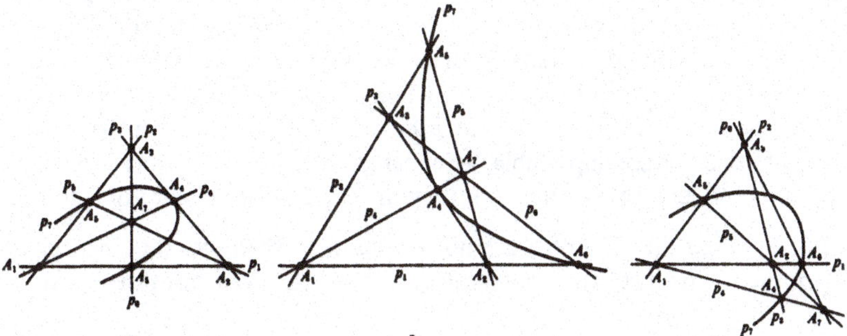

a. b. c.

Fig. 34

this it is sufficient to note that through the three points which lie on one of the lines there pass all the remaining lines (two through each of the points).

Let us start now with three points A_1, A_2, A_3 and the three lines, say p_1, p_2, p_3, which join these points in pairs (fig. 34). By the conditions of the problem, one of the seven given lines (other than the sides of this triangle) must pass through each vertex of the triangle $A_1A_2A_3$.

Suppose that p_4 passes through A_1, p_5 through A_2, and p_6 through A_3. Let p_4 meet p_3 at A_4, p_5 meet p_2 at A_5, and p_6 meet p_1 at A_6. If all the conditions are to be fulfilled, we must require in addition that the three lines p_4, p_5, p_6 meet at a point A_7 and that the points A_4, A_5, A_6 lie on a line p_7. We shall show that it is impossible to fulfill both conditions. In particular, we shall show that if p_4, p_5, p_6 meet in a point A_7, then A_4, A_5, A_6 cannot be collinear.

Any straight line which does not pass through a vertex of a triangle either cuts two sides of the triangle (fig. 35a) or fails to cut any of the sides (fig. 35b): it cannot cut just one side or cut all all three sides. If A_7 lies inside triangle $A_1A_2A_3$ (fig. 34a), then all the points A_4, A_5, A_6, will lie on

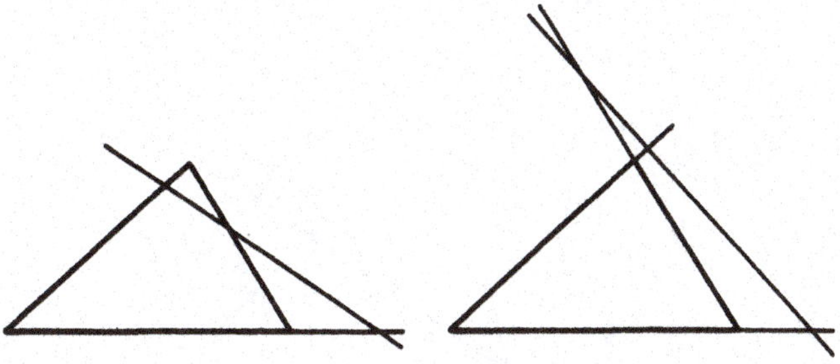

a. b.

Fig. 35

the sides of this triangle; and if it lies outside the triangle $A_1A_2A_3$ (fig. 34b and c), then just one of these points lies on a side of the triangle, and the others on the extensions of the other two sides. In the first case p_7 would have to cut all three sides of triangle $A_1A_2A_3$; in the second case it would have to cut just one side (and the extensions of the other two sides). Since both cases are impossible, we conclude that it is impossible to arrange seven points and seven lines to satisfy the conditions of the problem.

105. Suppose the lines of S are not all parallel; then two (and hence, by hypothesis, three) of them intersect in a point A. We shall prove that all

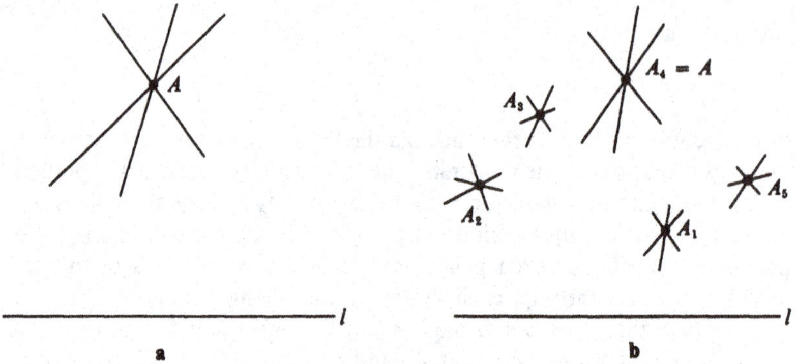

Fig. 36

the lines of S pass through A. To show this, we suppose there is a line l in S which does not pass through A (fig. 36a) and we derive a contradiction as follows. In addition to A there may be other intersection points of the lines of S which do not lie on l (fig. 36b). But since S is a finite set, there are only a finite number of such points, say $A_1, A_2, \ldots, A_k$. If d_i is the distance from A_i to l, we can choose the numbering so that d_1 is *minimal*, that is, $d_1 \leqq d_i$ for $i = 2, \ldots, k$.

By hypothesis there are at least three lines through A_1, and these lines intersect l in points P, Q, R, where Q lies between P and R (fig. 37).

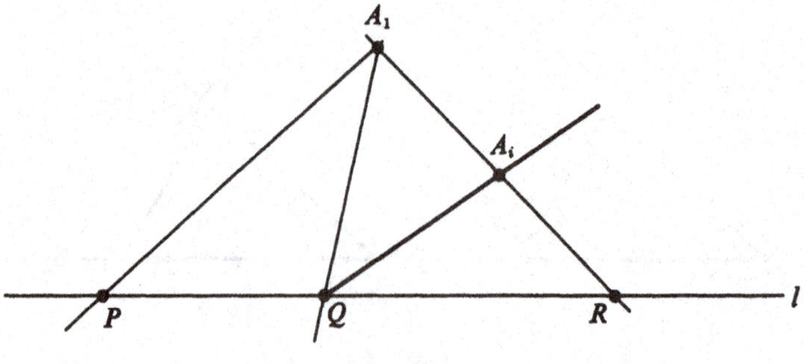

Fig. 37

By hypothesis S contains a line through Q other than l and A_1Q. This line intersects either A_1P or A_1R in a point A_i. But then $d_i < d_1$, a contradiction. This proves that all the lines of S pass through A.

106. As in problem 105, the proof is by contradiction. If A, B, C are any three points, let $d(A;\ BC)$ denote the distance from A to the line l containing B and C. Now suppose that not all points of S are collinear; then there exist three points A, B, C in S such that $d(A;\ BC) > 0$. There are only a finite number of such triplets A, B, C, for S is finite. Hence there is at least one such triplet such that $d(A;\ BC)$ is *minimal.*

By hypothesis the line l through B and C contains another point D of S (see fig. 38). The foot P of the perpendicular from A to l divides l

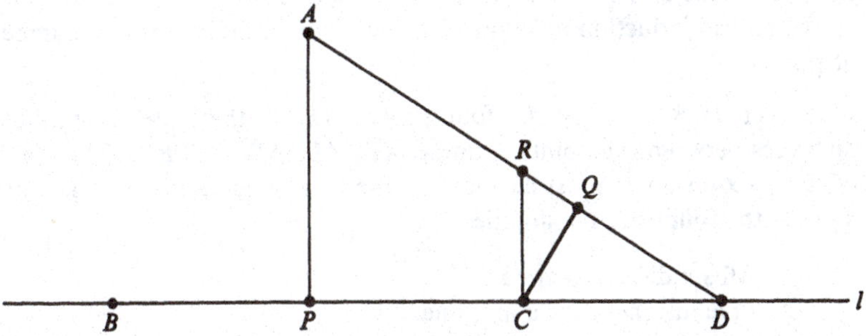

Fig. 38

into two halflines. Two of the three points B, C, D must lie on the same halfline (one of them may coincide with P). Let us say for definiteness that the situation is as shown in fig. 38, with C and D in the same halfline. Draw CQ perpendicular to AD and CR parallel to AP. Then $CQ < CR \leqq AP$, so that $d(C; AD) < d(A; BC)$. This contradicts the fact that $d(A; BC)$ was minimal, and the proof is complete.

Remark. The above proof is due to L. M. Kelly. For a historical account of this problem, together with other proofs, see Ref. [6].

The result can also be formulated as follows: Among the lines joining n points in the plane, not all collinear, there is at least one *ordinary* line, that is, a line containing only two of the points. In this formulation, one is naturally led to ask how many ordinary lines there must be. It has been shown that there must be at least $3n/7$ such lines; see Ref. [14].

107. The number n must be at least 3, since not all of the n points lie on a line. If $n = 3$ the theorem is clear: we need exactly three lines to join the three possible pairs of points. We now use mathematical induction. We suppose the theorem has been proved for n points and show that it also holds for $n + 1$ points. Consider all the lines joining our $n + 1$ points in pairs. According to problem 106 at least one of these lines must contain only two of our $n + 1$ points (or else all the points would lie on one line). Say these two points are A_n and A_{n+1}. If the n points $A_1, \ldots, A_n$ lie on a line, it is clear that the total number of lines is exactly $n + 1$ (the line on which the points $A_1, A_2, A_3, \ldots, A_n$ lie, and the n distinct lines joining these points to A_{n+1}). On the other hand, if the n points $A_1, A_2, A_3, \ldots, A_n$ do not lie on a line, then by the induction hypothesis there are at least n lines joining these points in pairs. Now draw lines joining A_{n+1} to $A_1, A_2, \ldots, A_n$. Since A_nA_{n+1} does not contain any of the other points $A_1, A_2, A_3, \ldots, A_{n-1}$, it is distinct from all the lines joining $A_1, A_2, A_3, \ldots, A_n$ in pairs. We have thus added at least one new line and therefore have a total of at least $n + 1$ distinct lines. Consequently, we have shown that if our theorem is true for n points, then it is also true for $n + 1$ points. By mathematical induction it follows that the theorem holds for any number of points.

108a. Let A, B, C, D be the four points. In all, there are six possible distances between the points, namely, AB, AC, AD, BC, BD, CD. Each of these six distances must have one of the two values a and b. There are *a priori* the following possibilities:

(1) All six distances are a.
(2) Five distances are a and one is b.
(3) Four distances are a and two are b.
(4) Three distances are a and three are b.

Let us examine these four cases separately.

Case 1. This case is impossible. For then the points A, B, C are the vertices of an equilateral triangle, and D, being equidistant from all of them, must lie at the center of the circumscribed circle. But then $AD < AB$, a contradiction.

Case 2. Three of the given points, say A, B, and C, must be vertices of an equilateral triangle of side a. The fourth point D must be at distance a from two of these vertices, say A and C, and at distance b from B.

So the points A, B, C, D are the vertices of a rhombus, one of whose diagonals, AC, is equal to its side (fig. 39a). It is easy to calculate that in this case $b = BD = a\sqrt{3}$.

Case 3. This case contains two subcases.

(i) Suppose the two segments of length b have a common endpoint, say D. Then the remaining points, A, B, C, form an equilateral triangle of side a. The point D is at distance b from two of the vertices, say A and C, and is at distance a from the third vertex B. Since $AD = CD$, D lies on the perpendicular bisector of AC; and since $BD = a = BA = BC$, D lies on the circle with center at B which passes through A and C. Hence there are two possible positions for D, as shown in fig. 39b and c. Thus

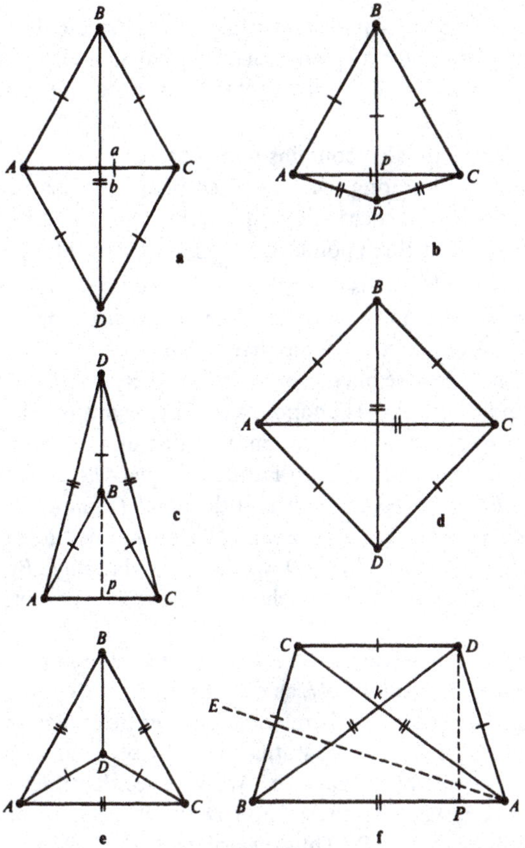

Fig. 39

A, B, C, D lie at the vertices of a deltoid,[1] two of whose sides are a and two b, and whose diagonals are each equal to a. It follows easily from the diagrams that in this case

$$b^2 = AP^2 + PD^2 = \left(\frac{a}{2}\right)^2 + \left(a \pm \frac{a\sqrt{3}}{2}\right)^2 = (2 \pm \sqrt{3})a^2,$$

that is, that

$$b = a\sqrt{2 + \sqrt{3}} \quad \text{or} \quad b = a\sqrt{2 - \sqrt{3}}.$$

(The two signs under the radical correspond to the two possibilities illustrated by figs. 39b and c.)

(ii) Suppose the two segments of length b do not have a common endpoint. Then the notation can be chosen so that $AC = BD = b$, and $AB = AD = BC = CD = a$. Since $BA = BC$ and $DA = DC$, both B and D lie on the perpendicular bisector of AD (fig. 39d). Moreover, since

[1] A deltoid is a quadrilateral with two pairs of equal adjacent sides.

$AB = AD$, B and D are equidistant from AC. Thus the diagonals of the quadrilateral $ABCD$ are perpendicular bisectors, and so $ABCD$ is a rhombus. Since $AC = BD$, the rhombus is a square, and therefore $b = a\sqrt{2}$.

Case 4. This case also contains two subcases.

(i) Suppose that among the four given points there are three (say A, B, C) which form the vertices of an equilateral triangle. Then D must be equidistant from these three points, and so lies at the center of the circumscribed circle (fig. 37c). It is clear that in this case $b = AD = a\sqrt{3}/3$.

(ii) Suppose that no three of our four points form the vertices of an equilateral triangle. Let us assume that $b > a$.

Among the three segments of length b there must be two with a common endpoint, for the six endpoints of these segments are all in the set $\{A, B, C, D\}$. Suppose such a common endpoint is A and that $AB = AC = b$. Since we have assumed that the triangle ABC is not equilateral, we must have $BC = a$. Notice now that the point D cannot be at the same distance from B and C (that is, cannot lie on the perpendicular AE to BC): for if it were, then either $BD = CD = a$ and triangle BCD would be equilateral, or $BD = CD = b$, which would make four of our six distances equal to b.

So the point D does not lie on AE, and we may suppose by symmetry that it lies on the same side of AE as C. Then $DB > DC$, so that we must have $DB = b$, $DC = a$. But there are altogether three segments of length a and three of length b; it follows that the only segment we have not yet considered, AD, must be of length a. We now see that the triangles ABC and BAD are congruent, so that C and D are at the same distance from AB and CD is parallel to AB. Thus the points A, B, C, D form an isosceles trapezoid whose shorter base is equal to the sides, and whose longer base is equal to the diagonals (fig. 39f).

Let P be the foot of the perpendicular from D to AB. Then $BD^2 = BP^2 + PD^2$ and $AD^2 = AP^2 + PD^2$, so that $BD^2 - AD^2 = BP^2 - AP^2 = (BP + AP)(BP - AP)$. Now $BD = b$, $AD = a$, $BP + AP = BA = b$, and $BP - AP = CD = a$. Thus $b^2 - a^2 = ba$. Dividing by a^2 and transposing, we obtain $(b/a)^2 - (b/a) - 1 = 0$. Hence $b/a = (1 + \sqrt{5})/2$ (since the other root of the quadratic equation $x^2 - x - 1 = 0$ is $(1 - \sqrt{5})/2$, which is negative).

Thus figs. 37a to f show all possible arrangements of four points in the plane such that the distance between any two of them is one of the values a and b. We see that such configurations are possible only if

$$b = a\sqrt{3}, \quad b = a\sqrt{2 + \sqrt{3}}, \quad b = a\sqrt{2 - \sqrt{3}},$$

$$b = a\sqrt{2}, \quad b = \frac{a\sqrt{3}}{3}, \quad \text{or} \quad b = a\,\frac{1 + \sqrt{5}}{2},$$

where b is the distance occurring less frequently, or if both distances occur the same number (three) of times, b is the longer. It is more convenient, however, to reformulate the result so that b is taken to be the longer of the two lengths in every case. Since $1/\sqrt{2-\sqrt{3}} = \sqrt{2+\sqrt{3}}$ and $1/(\sqrt{3}/3) = \sqrt{3}$, the following possibilities are the only ones that remain:

$$b = a\sqrt{3}, \quad b = a\sqrt{2+\sqrt{3}}, \quad b = a\sqrt{2}, \quad \text{and} \quad b = a\frac{1+\sqrt{5}}{2}.$$

b. Let us consider, in order, the possible values of n.

(1) $n = 2$. In this case there is only one distance: whatever the location of the two points in the plane, this distance assumes only one value a.

(2) $n = 3$. In this case there are three distances. If these distances are to assume only the two values a and b (say a twice and b once), the

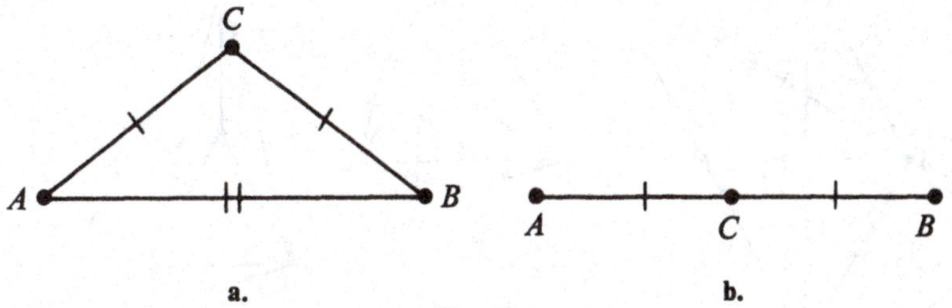

Fig. 40

points must be the vertices of an isosceles triangle with base b and side a (fig. 40a and b: the latter illustrates the degenerate case when the triangle reduces to a line segment). It is clear that a and b need only satisfy the inequality

$$a \geqq \frac{b}{2};$$

in particular, it is possible to have $a = b$.

(3) $n = 4$. This case was settled in part **a**.

(4) $n = 5$. If A, B, C, D are any four of the five points A, B, C, D, E, they satisfy the condition of part **a** and must therefore be arranged in one of the six configurations shown in fig. 39. Suppose, for example, that A, B, C, D are arranged as in fig. 39a. The distances between the four points A, B, C, E must assume the same two values a and $b = a\sqrt{3}$; it follows that these points must be arranged as in fig. 39a or e. But three of these four points, A, B, C, are the vertices of an equilateral triangle of side a, and in fig. 39c the only equilateral triangle has side $b > a$. So A, B, C, E must be situated as in fig. 39a. Since E does not coincide with D, the only

possible configuration is that illustrated in fig. 41a. But in this configuration $DE = 2a$ has length neither a nor b. So we have shown that it is impossible for four of our five points to be arranged as in fig. 39a.

Similarly, we can show that the points A, B, C, D cannot be situated as in fig. 39b or c, for if we assume that they are, we necessarily arrive at one of the configurations 39b or c for the five points A, B, C, D, E, and in each of these configurations the length of DE is different from a or b. If A, B, C, D are arranged as in fig. 39e, then so are A, B, C, E, and this is clearly impossible if D and E are to be distinct. Similarly, if A, B, C, D are as in 39d, E must coincide with D (for the four points A, B, C, E must also be at the vertices of a square).

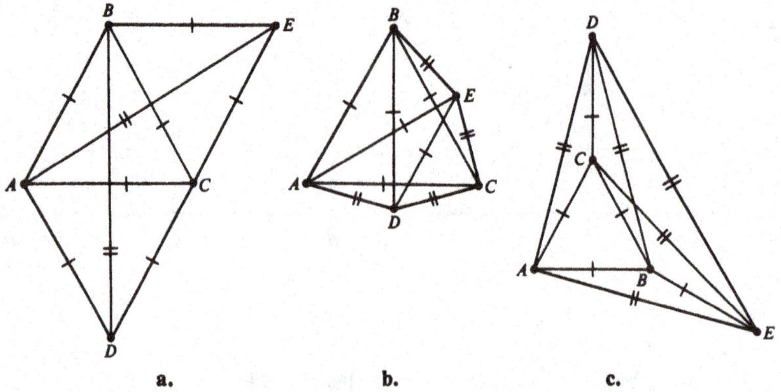

Fig. 41

Thus the only case remaining is the one in which A, B, C, D are at the vertices of the trapezoid illustrated in fig. 37f. In this case A, B, C, E must form a congruent trapezoid, and it is easy to see that the five points A, B, C, D, E must lie at the vertices of a regular pentagon (fig. 42).[2] This arrangement of the five points is therefore the only one satisfying the conditions of the problem: all the 10 distances between pairs of the five points are equal to one of the quantities a and $b = a(1 + \sqrt{5})/2$.

[2] Note that if we circumscribe a circle about the trapezoid $ABCE$ (fig. 37f), then the points A, B, C, D will be four of the vertices of a regular pentagon inscribed in this circle. To prove this, put $\angle DAB = \angle ADB = \alpha$. Then $\angle CDB = \angle CBD = \angle CDA - \angle BDA = (180° - \alpha) - \alpha$ (note that in the cyclic quadrilateral $ABCD$ the angles at B and D sum to 180°, and the angle at B is α by the congruence of triangles ABC and BAD) $= 180° - 2\alpha = \angle ABD$, and $\angle BCD = 180° - 2(180° - 2\alpha) = 4\alpha - 180°$. Since $\angle BCD + \angle CBA = 180°$, we have $5\alpha - 180° = 180°$; $\alpha = 2(180°/5) = 72°$. Now since $\angle ABD = \angle DBC = 180°/5$, it follows that AD, CD, and CB are three of the sides of a regular pentagon inscribed in the circle.

(5) $n \geqq 6$. By the result we have just obtained, any five of the given n points must lie at the vertices of a regular pentagon. But if $n \geqq 6$ points are all distinct, such an arrangement is clearly impossible. So there are no solutions to our problem for $n \geqq 6$.

We thus see that the problem has a solution only for $n = 2, 3, 4$, or 5.

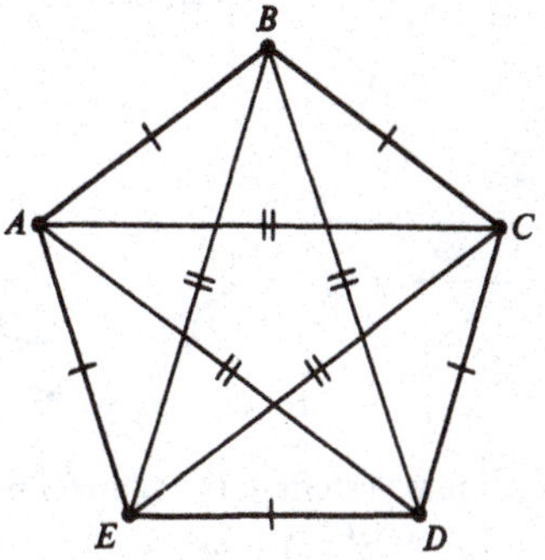

Fig. 42

109a. Let N be a given integer, greater than 2. We are required to find N points in the plane, not all collinear, such that the distance between any two of them is an integer.

We give two such arrangements.

First solution. It is easy to check directly that if

$$x = 2uv, \quad y = u^2 - v^2, \quad z = u^2 + v^2, \tag{1}$$

then

$$x^2 + y^2 = z^2. \tag{2}$$

By using these formulas[3] we can find $N - 2$ distinct triples of positive integers $(x_1,y_1,z_1), (x_2,y_2,z_2), \ldots, (x_{N-2},y_{N-2},z_{N-2})$ satisfying (2) and such that $x_1 = x_2 = \cdots = x_{N-2}$. For let k be any common multiple of the integers $1, 2, 3, \ldots, N - 2$ such that $k > (N - 2)^2$. To obtain our $N - 2$ triples we successively put

$$u_1 = k, \quad v_1 = 1; \qquad u_2 = \frac{k}{2}; \quad v_2 = 2,$$

$$u_3 = \frac{k}{3}, \quad v_3 = 3\,; \ldots; \qquad u_{N-2} = \frac{k}{N-2}, \quad v_{N-2} = N - 2$$

in formula (1). This gives $N - 2$ triples (x_i,y_i,z_i). Clearly $x_1 = x_2 = \cdots = x_{N-2} = 2k$, and $y_1 > y_2 > \cdots > y_{N-2}$. Moreover, $y_{N-2} > 0$, since

$$y_{N-2} = \left(\frac{k}{N-2}\right)^2 - (N-2)^2 > \left(\frac{(N-2)^2}{N-2}\right)^2 - (N-2)^2 = 0.$$

[3] Equations (1) are the well-known formulas for the lengths of the sides of a right-angled triangle, when all these lengths are integers. See, for example, Hardy and Wright, Ref. [9].

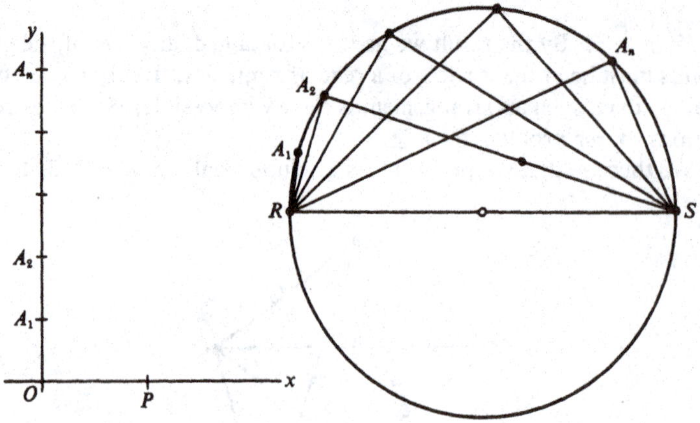

a Fig. 43 b

Now choose a coordinate system in the plane, and mark off $n = N - 2$ points $A_1, A_2, A_3, \ldots, A_n$ on the y axis at distances $y_1, y_2, y_3, \ldots, y_n$ from O. Let P be the point $(2k,0)$ (fig. 43a). It is easy to see that the N points $O, P, A_1, A_2, A_3, \ldots, A_{N-2}$ satisfy the conditions of the problem, that is, that the distance between any two of them is an integer.

Remark. Instead of marking off $N - 1$ points $O, A_1A_2, \ldots, A_{N-2}$ on Oy and a single point P on Ox, we can take only $N - 2$ points on Oy, and for our Nth point take $P' = (-2k, 0)$. We could also take only half our points on the positive y axis, and choose the remainder so that the whole configuration is symmetric about O.

Second solution. We see from equations (1) and (2) that

$$\left(\frac{2uv}{u^2 + v^2}\right)^2 + \left(\frac{u^2 - v^2}{u^2 + v^2}\right)^2 = 1. \tag{3}$$

Using (3) we can find N points $R, S, A_1, A_2, \ldots, A_{N-2} = A_n$ on the circumference of a circle with diameter $RS = 1$ so that all the distances $A_1R, A_2R, \ldots, A_nR, A_1S, A_2S, \ldots, A_nS$ are rational numbers: for this choose $n = N - 2$ pairs of rational numbers $(u_1,v_1), (u_2,v_2), \ldots, (u_n,v_n)$ so that the ratios u_i/v_i are distinct and >1. Take for $A_1, A_2, \ldots, A_n$ the points for which $A_iR = (2u_iv_i)/(u_i^2 + v_i^2)$; $i = 1, 2, \ldots, n$; note that in this case

$$A_iS = \frac{u_i^2 - v_i^2}{u_i^2 + v_i^2}$$

(fig. 43b).

We prove now that the distance between any two of these points is also rational. Since the diameter of the circle is 1, we have

$$\begin{aligned} A_1A_2 &= \sin A_1RA_2 = \sin (A_1RS - A_2RS) \\ &= \sin A_1RS \cos A_2RS - \sin A_2RS \cos A_1RS \\ &= A_1S \cdot A_2R - A_2S \cdot A_1R; \end{aligned}$$

from which it follows at once that A_1A_2 is rational. By an analogous argument each of the distances A_iA_j $(i, j = 1, 2, \ldots, n)$ is rational.

Suppose k is any common multiple (for instance, the least common multiple) of the denominators of the fractions representing the distances between all possible pairs of points selected from $R, S, A_1, A_2, \ldots, A_n$. Then by drawing a figure similar to the one we have constructed, but magnified by a factor of k, we obtain N points, any two of which are an integral distance apart. All these points lie on a circle of radius k, but not on a straight line.

b. We must show that an infinite number of points, not all lying on a straight line, cannot have the property that the distance between any two of them is an integer. To do so it is sufficient to prove that given three

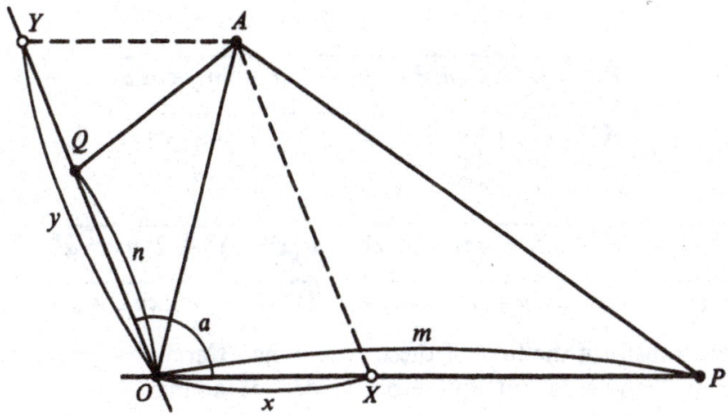

Fig. 44

noncollinear points O, P, Q, there exist only a finite number of points at integral distances from all of them.

Suppose $PO = m$ and $QO = n$, where m and n are integers. We denote by A an arbitrary point lying at an integral distance from the three points O, P, Q (fig. 44). Since the difference between the lengths of two sides of a triangle does not exceed the length of the third side,

$$|AP - AO| \leqq PO = m \quad \text{and} \quad |AQ - AO| \leqq QO = n,$$

equality being attained when the triangle APO (or AQO) degenerates into a segment of a straight line. Since the distances AO, AP, and AQ are integers, the differences $AP - AO$ and $AQ - AO$ are also integers (possibly negative). From the given inequalities, the difference $AP - AO$ can assume only the $2m + 1$ distinct values $m, m - 1, \ldots, 1, 0, -1, \ldots, -m + 1, -m$, and $AQ - AO$ the $2n + 1$ values $n, n - 1, \ldots, 1, 0, -1, \ldots, -n + 1, -n$.

We now show that for given values k and l of the two differences $AP - AO$ and $AQ - AO$, there can be at most two positions for the point

A; it then follows that there cannot be more than $2(2m + 1)(2n + 1)$ distinct positions for A, such that the distance from A to each of the points O, P, Q is integral. This will conclude our proof.

Suppose then that

$$AP - AO = k \quad \text{and} \quad AQ - AO = l.$$

Draw lines AX and AY which pass through A and are parallel to OQ and OP, respectively (fig. 44). Let x and y be the lengths $OX = YA$ and $OY = XA$, and let α be the angle QOP. We note that $\alpha \neq 0°$; $\alpha \neq 180°$, since O, P, Q are not collinear. On applying the law of cosines to the triangles OAX, PAX, and QAY, and noting that $\cos AXO = \cos(180° - \alpha) = \cos \alpha$, we find that

$$AO = \sqrt{x^2 + y^2 + 2xy \cos \alpha}$$

$$PA = \sqrt{(x - m)^2 + y^2 + 2(x - m)y \cos \alpha}$$

$$AQ = \sqrt{(y - n)^2 + x^2 + 2(y - n)x \cos \alpha}.$$

Since $AP - AO = k$ and $AQ - AO = l$,

$$\sqrt{[(x - m)^2 + y^2 + 2(x - m)y \cos \alpha]} - \sqrt{[x^2 + y^2 + 2xy \cos \alpha]} = k, \quad (1)$$

$$\sqrt{[x^2 + (y - n)^2 + 2x(y - n) \cos \alpha]} - \sqrt{[x^2 + y^2 + 2xy \cos \alpha} = l. \quad (2)$$

We shall transform the first of these equations. Carry the second radical to the right-hand side and square both sides, obtaining

$$(x - m)^2 + y^2 + 2(x - m)y \cos \alpha$$
$$= x^2 + y^2 + 2xy \cos \alpha + 2k\sqrt{(x^2 + y^2 + 2xy \cos \alpha)} + k^2$$

or

$$x^2 - 2xm + m^2 + y^2 + 2xy \cos \alpha - 2my \cos \alpha$$
$$= x^2 + y^2 + 2xy \cos \alpha + 2k\sqrt{(x^2 + y^2 + 2xy \cos \alpha)} + k^2,$$

or

$$-2xm - 2ym \cos \alpha + m^2 - k^2 = 2k\sqrt{(x^2 + y^2 + 2xy \cos \alpha)}.$$

In exactly the same way the second equation yields

$$-2yn - 2xn \cos \alpha + n^2 - l^2 = 2l\sqrt{(x^2 + y^2 + 2xy \cos \alpha)}.$$

On comparing these last two equations, we find that

$$l(-2xm - 2ym \cos \alpha + m^2 - k^2) = k(-2yn - 2xn \cos \alpha + n^2 - l^2),$$

or

$$x(-2ml + 2nk \cos \alpha) + y(-2ml \cos \alpha + 2nk)$$
$$= k(n^2 - l^2) - l(m^2 - k^2), \quad (3)$$

and on squaring each of them, we obtain

$$(-2xm - 2ym\cos\alpha + m^2 - k^2)^2 = 4k^2(x^2 + y^2 + 2xy\cos\alpha), \quad (4)$$

and

$$(-2yn - 2xn\cos\alpha + n^2 - l^2)^2 = 4l^2(x^2 + y^2 + 2xy\cos\alpha). \quad (5)$$

If the coefficients of x and y in (3) are not both zero, we may solve for one of them and substitute in (4) or (5): we obtain a quadratic, which cannot have more than two distinct real roots. We thus obtain at most two solutions [that is, pairs (x,y)] for the equations, and each pair determines uniquely the position of the point A [in effect, the pair (x,y) is the expression for A in nonrectangular coordinates referred to axes OP and OQ]. Hence in this case there are at most two positions for A, and the theorem is proved.

If the coefficients of both x and y in (3) are zero, we have

$$m^2l^2 = mlnk\cos\alpha = n^2k^2.$$

Thus $ml = nk$ or $-nk$ and either $ml = nk = 0$ or $\cos\alpha = \pm 1$. But $\cos\alpha = \pm 1$ is impossible, as we have noted, and the remaining case reduces to

$$l = 0 = k.$$

The equations (4) and (5) now reduce to

$$2x + 2y\cos\alpha = m$$

$$2y + 2x\cos\alpha = n$$

and these may be solved uniquely for x and y, since $\cos^2\alpha \neq 1$. Thus in this case there is only one possible position for the point A.

We have thus completed an algebraic proof that the number of points A for which $AP - AO = k$ and $AQ - AO = l$ (where O, P, Q are given noncollinear points of the plane, and k and l are given real numbers) is at most two. To explain this result geometrically, it is natural to consider the locus of those points of the plane whose distances from two given points have a constant difference. This locus is studied in detail in analytic geometry: it is a branch of the curve known as a hyperbola (a special case of which is the graph of the function $y = 1/x$) and has the form given in fig. 45.[4]

From the shape of such curves it is intuitively clear that two of them cannot intersect in more than two points (see fig. 45b), and it is just this fact which was proved rigorously above.

[4] In figs. 45a and b the second branches of the hyperbolas are shown as dotted lines. It is customary to give the name *hyperbola*, not to the locus of points from which the distances to two given fixed points have a constant difference but to the locus of points for which the difference of these distances has a constant *absolute value*. The hyperbola thus has two branches: one consisting of the points M for which $MA - MB = a$, and the other of the points N for which $NB - NA = a$.

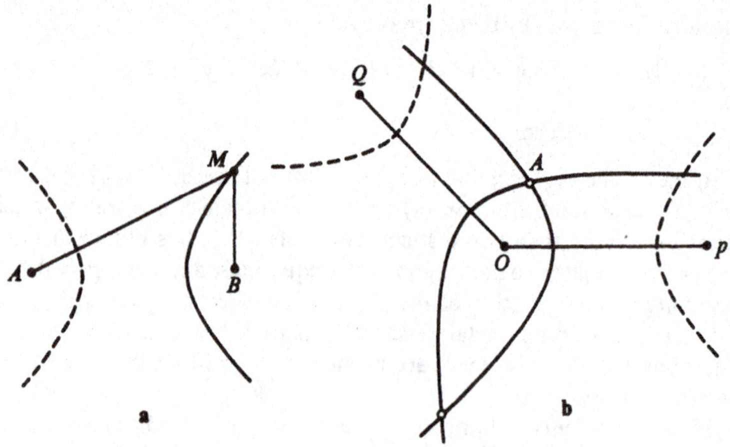

Fig. 45

We note that if O, P, Q are collinear, the proof breaks down; then the very definition of the quantities x and y is impossible. Geometric considerations in this case do not lead to a solution either: if O, P, Q are collinear, the hyperbolas can degenerate into rays of the line OPQ (this will happen, for example, when $k = OP$ and $l = OQ$) and coincide with each other along a whole halfline. Of course, this corresponds to the fact that it *is* possible to find an infinite number of points on a line satisfying the conditions of the problem: we need merely choose all the points lying at an integral distance from some fixed point of the line.

II. LATTICES OF POINTS IN THE PLANE

110. For each lattice point L, let S_L be the lattice square whose lower left corner is L (fig. 46).[5] The set M is decomposed into pieces $M \cap S_L$ by the squares S_L. Translate each square S_L along the segment LO so that the squares are all superimposed on S_O. The pieces $M \cap S_L$ are thereby translated onto subsets T_L of S_O. Since the total area of these pieces is, by hypothesis, greater than 1, at least two of them must intersect. Suppose therefore that (x,y) is in both T_P and T_Q, where $P \neq Q$. If $P = (a,b)$ and $Q = (c,d)$, then the point $(x_1,y_1) = (x + a, y + b)$ is in $M \cap S_P$, while $(x_2,y_2) = (x + c, y + d)$ is in $M \cap S_Q$. Thus both (x_1,y_1) and (x_2,y_2) are

[5] More precisely, let S_L be the set of all points (x,y) such that $m \leqq x < m + 1$ and $n \leqq y < n + 1$, where $L = (m,n)$. The purpose of this convention is to insure that the squares S_L cover the plane without overlapping.

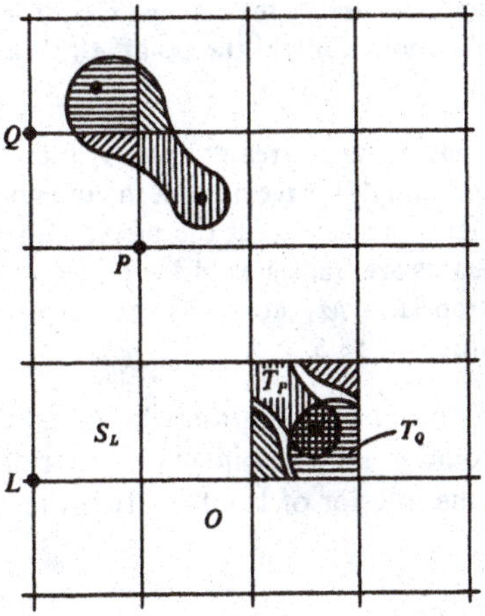

Fig. 46

in M. Furthermore, $x_2 - x_1$ and $y_2 - y_1$ are integers, since $x_2 - x_1 = c - a$ and $y_2 - y_1 = d - b$.

Remark. The result of this problem may be expressed in a different form as follows. Let L be a lattice point, and suppose each point A of M is translated to the point A' such that AA' is parallel and equal to OL (fig. 47). We then say that M has been given the translation OL, and we denote the image of M under this translation by M_L. It is easily seen from the figure that if $A = (x,y)$ and and $L = (m,n)$, then $A' = (x + m, y + n)$.

Now suppose M has an area greater than 1. By problem 110, M contains two distinct points of the form (x,y) and $(x + m, y + n)$, where m and n are

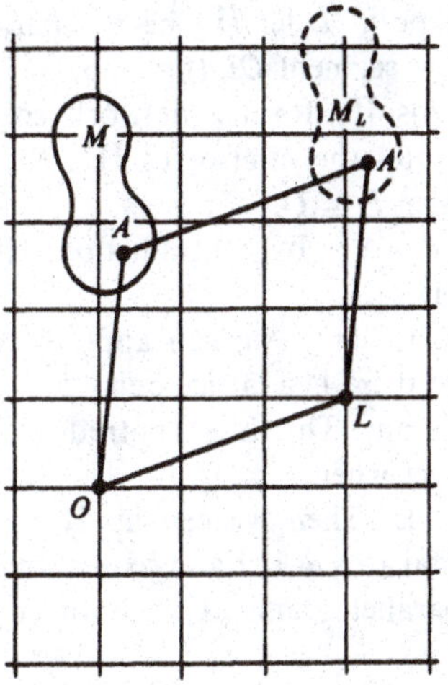

Fig. 47

integers. If $A = (x,y)$ and $L = (m,n)$, then $A' = (x + m, y + n)$ is by definition in M_L. Thus M and M_L both contain the point A'. We state this result as a theorem.

Theorem 1. If M has an area greater than 1, then there is some lattice point $L \neq 0$ such that M and M_L have a point in common.

We also note for later purposes that the pieces into which M_L is decomposed by the lattice squares are translates of the pieces T_P. Hence we have

Theorem 2. The translates M_L cover the entire plane if and only if the pieces T_P cover the entire square S_O.

111a. Let $\Pi = OABC$ be a parallelogram whose vertices are lattice points and which contains no other lattice points in its interior or on its boundary (fig. 48a). Let Π^i be the interior of Π; then Π is the union of Π^i and the

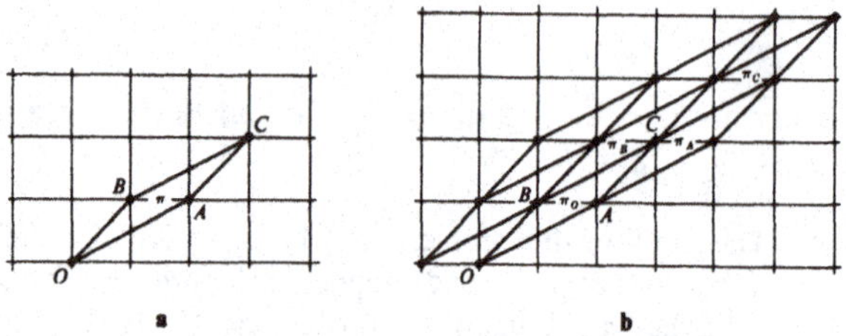

Fig. 48

four edges OA, AC, CB, BO. Clearly Π and Π^i have the same area α; the problem is to prove that $\alpha = 1$.

For each lattice point L let Π_L be the parallelogram obtained by translating Π along the segment OL (fig. 48b).

The parallelograms Π_L form a network covering the entire plane. Moreover, if Π_L^i denotes the interior of Π_L, then Π_P^i and Π_Q^i have no points in common when $P \neq Q$.

The proof that $\alpha = 1$ is by contradiction. If $\alpha \neq 1$, we must have either $\alpha > 1$ or $\alpha < 1$.

(1) Suppose $\alpha > 1$. Then we can apply theorem 1, with $M = \Pi^i$. The theorem says that there is a lattice point $L \neq O$ such that Π^i and Π_L^i have a point in common. This is a contradiction to the fact that Π_P^i and Π_Q^i do not intersect when $P \neq Q$.

(2) Suppose $\alpha < 1$. Then we can apply theorem 2 with $M = \Pi$. The pieces T_P have total area $\alpha < 1$ and so cannot cover all of S_O. Hence, by theorem 2, the parallelograms Π_L cannot cover the entire plane, a contradiction.

These contradictions prove that $\alpha = 1$ as required.

b. It follows from part **a** that the formula (3) $A = i + b/2 - 1$ holds when Π is a parallelogram containing no lattice points other than its vertices. For then $i = 0$, $b = 4$, and $i + b/2 - 1 = \frac{4}{2} - 1 = 1 = A$.

Next, suppose Π is an "empty" triangle, that is, a triangle containing no lattice points other than its vertices. Then $i = 0$, $b = 3$, and $i + b/2 - 1 = \frac{1}{2}$. Thus in order to verify (3) we must show that Π has area $\frac{1}{2}$. To prove this we extend Π to a parallelogram by adjoining a triangle Π' symmetric to it with respect to the center O of one of its sides (fig. 47). Since O is a center of symmetry of the lattice, Π' contains no lattice points

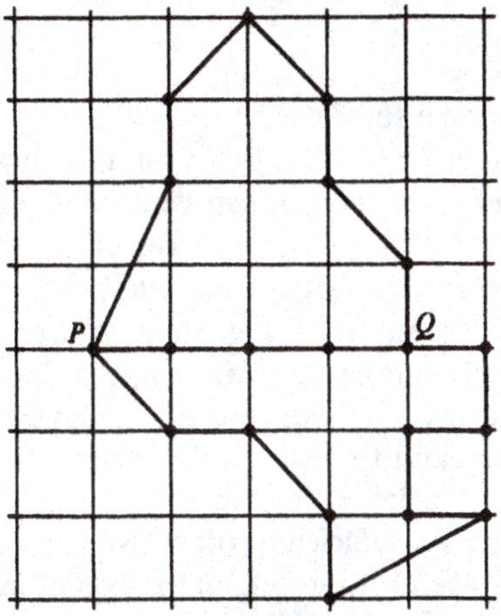

Fig. 49

other than its vertices. Hence, by the result of part **a**, the parallelogram we have constructed has area 1; therefore Π has area $\frac{1}{2}$.

Consider now two polygons Π_1 and Π_2 having all their vertices at lattice points and adjoining each other along the common side PQ (fig. 49).

Suppose we already know that the formula (3) holds for each of these polygons. We shall then show that it also holds for the polygon $\Pi = \Pi_1 \cup \Pi_2$. Let $A, A_1, A_2, i, i_1, i_2, b, b_1, b_2$ be the areas, numbers of interior lattice points, and boundary lattice points of Π, Π_1, and Π_2, respectively. By hypothesis

$$A_1 = i_1 + \frac{b_1}{2} - 1$$

and

$$A_2 = i_2 + \frac{b_2}{2} - 1.$$

Denote by k the number of lattice points on the segment PQ, including the two endpoints. Then it is clear from the figure that

$$A = A_1 + A_2,$$
$$i = i_1 + i_2 + (k - 2),$$

and

$$b = (b_1 - k) + (b_2 - k) + 2.$$

(The last term +2 takes account of the points P and Q.) Therefore

$$\begin{aligned} A = A_1 + A_2 &= i_1 + \frac{b_1}{2} - 1 + i_2 + \frac{b_2}{2} - 1 \\ &= (i_1 + i_2 + k - 2) + \frac{1}{2}(b_1 + b_2 - 2k + 2) - 1 \\ &= i + \frac{b}{2} - 1, \end{aligned}$$

which is what we wanted to prove.

It is now easy to show that (3) holds for all polygons with vertices at lattice points. Every such polygon can be decomposed into triangles by drawing suitable diagonals[6]; all the vertices of these triangles are lattice points, since they are vertices of the original polygon. Each triangle Π which is not already empty (that is, contains at least one lattice point P other than its vertices) can be split into empty triangles. For if P is an interior lattice point of Π, we join it to the three vertices, whereas if P is on an edge of Π, we join it to the opposite vertex. In either case we have split Π into smaller triangles, each containing fewer lattice points than Π. By repeating this process sufficiently often, we eventually decompose all the triangles, and hence also the original polygon, into empty triangles. As we have seen, the formula (3) holds for the empty triangles. By repeatedly applying the fact that (3) holds for the union of two polygons if it holds for each of them, we see that (3) holds for the original polygon.

Remark. For any polygon Π we have $b \geqq 3$ (since Π has at least three vertices), and $i \geqq 0$. Hence $A = i + b/2 - 1 \geqq \frac{3}{2} - 1 = \frac{1}{2}$. Thus any polygon whose vertices are lattice points has area at least $\frac{1}{2}$. Equality holds if and only if Π is an empty triangle.

112. If $A = (x,y)$ is any point of K, let $A' = (x/2,y/2)$. It is easily seen that A' lies on the segment OA, and $\overline{OA'} = \frac{1}{2}\overline{OA}$ (fig. 50).

Applying the transformation $A \to A'$ to every point of K, we obtain a set K' which is similar to K in the ratio 1:2. Since the area of a set is proportional to the square of its linear dimensions, area $(K') = \frac{1}{4}$ area (K). By hypothesis, area $(K) > 4$, and therefore area $(K') > 1$. By Blichfeldt's lemma (problem 110), K' contains two points (x_1,y_1) and (x_2,y_2) such that $x_2 - x_1$ and $y_2 - y_1$ are integers. By the construction of K', the points $A = (2x_1,2y_1)$ and $B = (2x_2,2y_2)$ are in K. Since K is symmetric about O, it also contains the point $C = (-2x_1,-2y_1)$, which is the mirror image of A in O. Since K is convex it contains the midpoint P of the segment CB.

[6] See, for example, Knopp, Ref. [15].

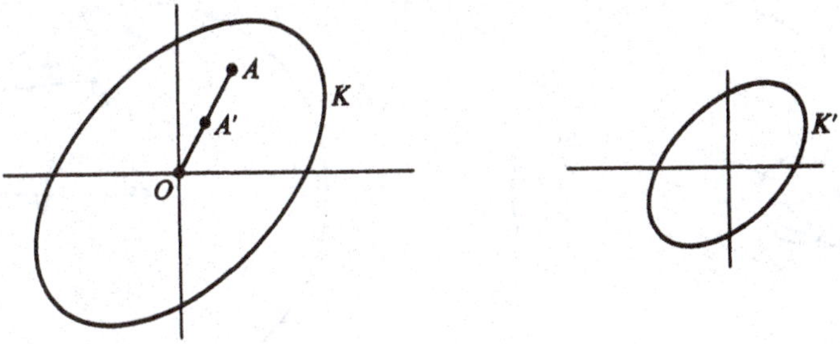

Fig. 50

The coordinates of P are

$$\left(\frac{-2x_1 + 2x_2}{2}, \frac{-2y_1 + 2y_2}{2}\right) = (x_2 - x_1, y_2 - y_1).$$

Since $x_2 - x_1$ and $y_2 - y_1$ are integers, P is a lattice point.

Remarks. (1) Since K is symmetric about O, it contains $Q = (x_1 - x_2, y_1 - y_2)$, the mirror image of P in the origin. Thus K contains at least two lattice points other than O.

(2) If K is convex, symmetric about O, and has area 4, it need not contain any lattice point other than O. Consider, for example, the square consisting of all points (x,y) with $|x| < 1$, $|y| < 1$. However, there is always a lattice point $P \neq 0$ either inside K or on its boundary. For if there were no such point, we could magnify K slightly about O and obtain a set L which would still contain no lattice points except O. This is a contradiction to Minkowski's theorem, since L has area >4.

113. We show first that if the radius ρ of the trees is greater than $\frac{1}{50}$, then the view from the center will be blocked in every direction. Through the center O draw an arbitrary line MN intersecting the boundary of the orchard in the points M and N. We assert that there is no gap in the trees in either of the directions OM and ON. To prove this, draw tangents to the boundary of the orchard at M and N, and lines AD and BC equal and parallel to MN and at distance ρ from it (fig. 51a). We obtain a rectangle $ABCD$ of area $AB \cdot MN = 100 \cdot 2\rho = 4 \cdot 50\rho$ and so greater than 4 (since $\rho > \frac{1}{50}$). By Minkowski's theorem (problem 112) it follows that there are at least two lattice points P and Q inside the rectangle which are symmetric with respect to O. (See remarks at the end of the solution to problem 112.) The trees of radius ρ planted at P and Q intersect the rays OM and ON, so that an observer at the center cannot see out of the orchard in either of these directions.

We must now show that if $\rho < 1/\sqrt{2501}$, it is possible to see out of the orchard from O. Let R be the point whose coordinates are (50,1). (See fig. 51b.) We will prove that the segment OR does not intersect any trees.

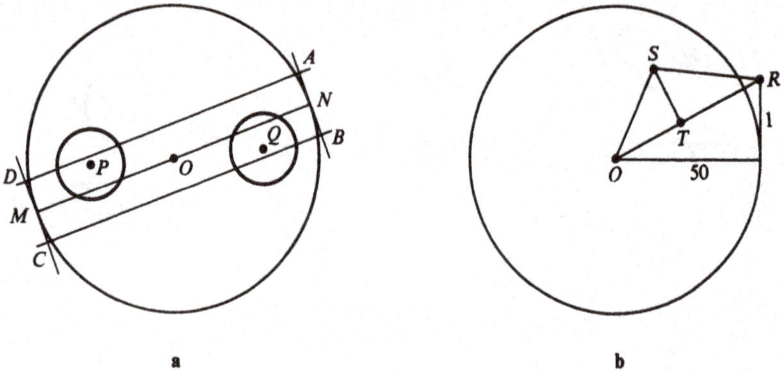

Fig. 51

Clearly there are no lattice points on OR other than the endpoints O and R. The length of OR is $\sqrt{50^2 + 1^2} = \sqrt{2501}$. Now let S be any lattice point in the orchard. By the remark made in the solution of problem 111**b**, the area of the triangle ORS is at least $\frac{1}{2}$.

If T is the foot of the perpendicular from S to OR, then the area of ORS is $\frac{1}{2}\overline{OR} \cdot \overline{ST}$. Therefore $\frac{1}{2}\overline{OR} \cdot \overline{ST} \geqq \frac{1}{2}$, which implies that $\overline{ST} \geqq 1/\overline{OR} = 1/\sqrt{2501} > \rho$. This means that the tree of radius ρ with center at S does not intersect OR. Thus it is possible to see out of the orchard by looking toward R.

III. TOPOLOGY

114. To begin with, it is clear that the two regions into which the plane is divided by a single straight line can be colored according to the conditions of the problem (fig. 52). We shall now show that if the regions into which

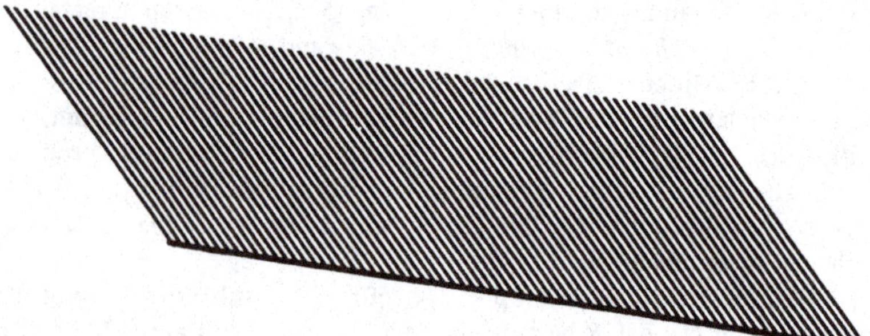

Fig. 52

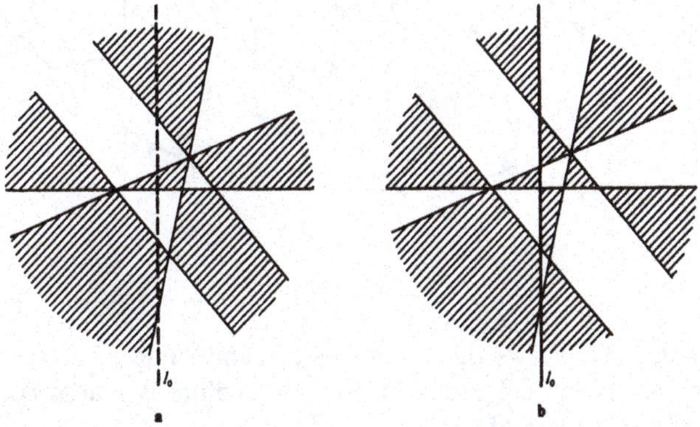

Fig. 53

the plane is divided by n lines can be properly colored, then so can the regions into which it is divided by $n + 1$ lines. The result then follows by mathematical induction.

Suppose that we are given $n + 1$ lines. Throw one of them away. The remaining n lines divide the plane into areas which by the induction hypothesis can be colored to suit the conditions of the problem (fig. 53a). Let us now reintroduce the $(n + 1)$st line and change all the colors on one side of it, replacing black by white and white by black, while leaving all the colors on the other side unchanged (fig. 53b). The plane is now properly colored. For if two of the regions into which the plane is divided by the $n + 1$ lines adjoin each other along one of the first n lines, then they must have different colors. (The reason is that they had opposite colors before, which have either remained unchanged or both been changed.) If two regions touch along the $(n + 1)$st line, then before this line was reintroduced they were part of the same region and thus had the same color. But we changed the colors on one side of the line and not on the other, so that now they have opposite colors. Thus the coloring we have constructed satisfies the conditions of the problem, and the theorem is proved.

115a. *First solution.* Let 1, 2, 3 denote three colors. Since at most two lines meet at any node, each connected part of the network is either an open or a closed polygonal path (fig. 54a). Any open path can be colored by painting its lines alternately 1 and 2.

The same rule applies to any closed path with an even number of sides. If P is a closed path with an odd number of sides, say $2m + 1$, then we can color $2m$ of the sides by alternating the colors 1 and 2. Then the remaining side can be colored 3.

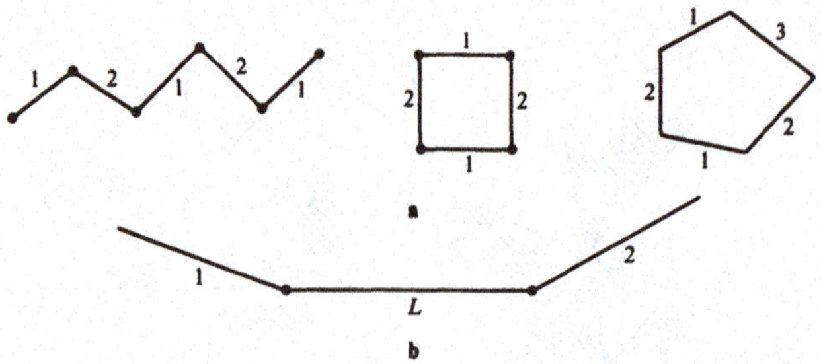

Fig. 54

This analysis shows that the necessary and sufficient condition that our network can be colored with only two colors is that it contain no closed paths with an odd number of sides.

Second solution. Choose some line of the network and color it 1. Then choose another line and color it 2. Proceed in this way, coloring one line after another. Suppose that at the *n*th stage of this process we have chosen a line L to be colored. By hypothesis there is at most one other line through each endpoint of L (fig. 54b). These other lines may have already been painted, but since there are altogether three colors, there is at least one color left for L. Thus the process can be continued until the entire network is colored.

b. Denote the four available colors by 1, 2, 3, 4. Figure 55a illustrates an example where all four colors must be used, since every line is adjacent to every other line.

We must now prove that four colors are always sufficient. As in the second solution to part a, start coloring one line after another. Suppose

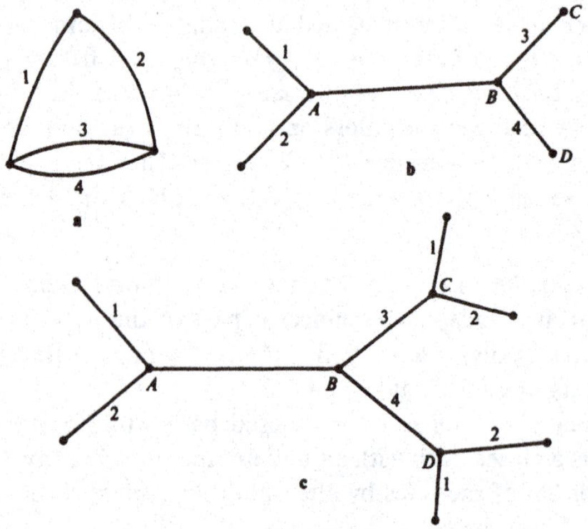

Fig. 55

that at the nth stage of the process we have chosen the line AB to be colored. By hypothesis there are at most two other lines through each of the points A and B. Hence there is a color available for AB unless these lines have already been colored 1, 2, 3, and 4 as shown in fig. 53b.

Unless there are lines through C colored 1 and 2, we can change the color of BC to 1 or 2, thereby releasing 3 for use on AB. Similarly, we are in good shape unless there are lines through D colored 1 and 2. Hence we may suppose that the situation is as shown in fig. 55c.

Now let P be the longest possible path starting at A, with the property that its edges are alternately colored 1 and 3. There are three possibilities:

(1) P ends at B.
(2) P ends at D.
(3) P does not end at B or D.

These three cases are illustrated in fig. 56.

In cases (2) and (3) we interchange the colors 1 and 3 along the path P. The result of this interchange is that adjacent lines are still colored differently, but now AB can be colored 1.

In case (1) let Q be the longest path through D whose edges are alternately colored 1 and 3. This path cannot end at any of the lettered vertices. We interchange the colors 1 and 3 on the edges of Q. Then the color of BD can be changed to 1, thus releasing 4 for use on AB.

Hence in all cases we can color AB, and so the process can be continued until the entire network is colored.

Remark. For another solution, see Ref. [13]. In the same way it can be shown that if at most $m = 2k$ lines meet at each node, then the network can be painted with $3k$ colors, whereas if $m = 2k + 1$, then $3k + 1$ colors are enough.

116a. We will prove that the number of triangles whose vertices are labeled 1, 2, 3 is odd (and therefore >0). Let the small triangles be $T_1, T_2, \ldots, T_n$, and denote by a_i the number of "12-edges" of T_i (that is, edges whose endpoints are labeled 1, 2. If the vertices of T_i are 1, 2, 3, then $a_i = 1$; if the vertices of T_i are 1, 1, 2 or 1, 2, 2, then $a_i = 2$; otherwise $a_i = 0$. The idea of the proof is to evaluate the sum $a_1 + a_2 + \cdots + a_n$ in two different ways.

(1) Let X denote the number of triangles T_i whose vertices are labeled 1, 2, 3, and let Y be the number whose vertices are labeled 1, 1, 2 or 1, 2, 2. Then we have

$$a_1 + a_2 + \cdots + a_n = X + 2Y.$$

(2) Let U be the number of 12-edges inside T, and let V be the number of 12-edges on the boundary of T. Every interior 12-edge lies in two triangles T_i and is therefore counted twice in the sum $a_1 + a_2 + \cdots + a_n$.

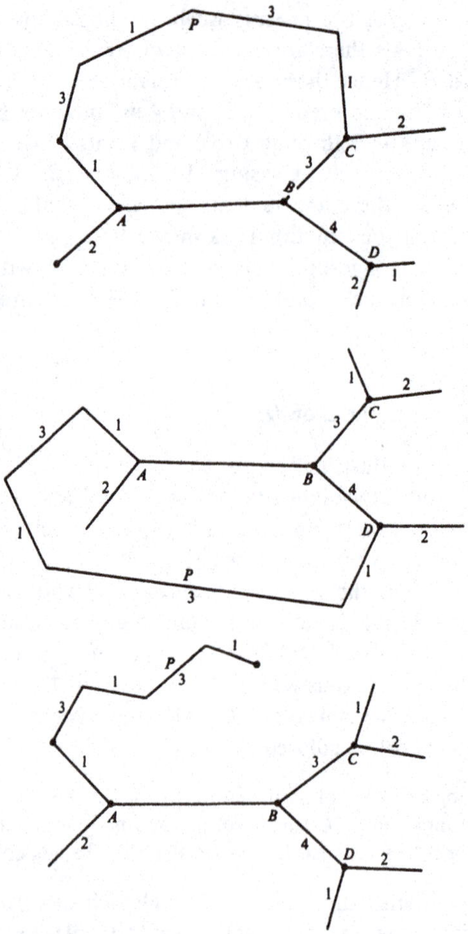

Fig. 56

But the 12-edges on the boundary of T lie on only one triangle T_i and so are counted only once in the sum. Hence

$$a_1 + a_2 + \cdots + a_n = 2U + V.$$

Comparing the two expressions, we get

$$X + 2Y = 2U + V.$$

It follows from this formula that X is even if V is even and odd if V is odd. It is therefore sufficient to show that V is always odd.

According to the conditions of the problem the side 13 cannot include any vertices numbered 2, and the side 23 cannot contain any vertices numbered 1, so that segments 12 can occur only on the side 12 of the large triangle T. Let us look at all such segments in the order in which they occur as we move from the endpoint 1 toward the endpoint 2. We first encounter a number (at least one) of vertices marked 1 (fig. 57). When we first pass a segment 12 we are at a vertex labeled 2. A number of 2's may now follow (that is, a number of segments 22): only after we have passed a segment 21 do we reach a vertex labeled 1. Passing the next segment 12 we again find ourselves at a vertex labeled 2; passing the next segment 21 we arrive at a 1 and so on.

Thus after an odd number of segments 12 we arrive at vertices labeled 2, and after an even number of segments at vertices labeled 1. But since the last vertex we come to is the vertex 2 of T, the total number of segments 12 lying on the side 12 of T must be odd. Thus the theorem is proved.

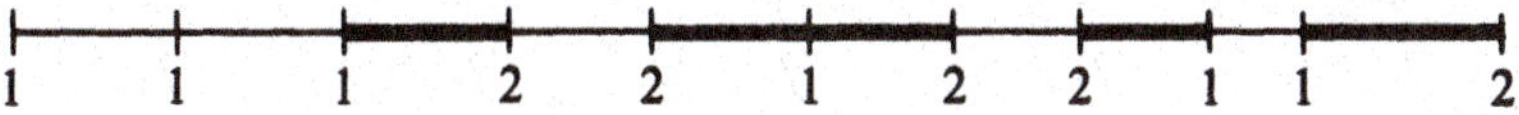

Fig. 57

Remark. This result can be put in a more precise form. Let us distinguish between the 123-triangles for which the vertices labeled 1, 2, 3 go around the triangle in the same order in which they go around T, and those for which they go around the opposite way. It can be shown that the number of triangles of the first type is always exactly one larger than the number of the second type; the result that the total number of triangles numbered 123 is odd follows from this. The proof of this assertion is similar to the solution we have given of the main problem and is left to the reader.

b. For the formulation of the theorem see the hints at the end of the book. The proof is similar to that of part **a**. We count in two different ways the number of 123-faces of the small tetrahedra. Each small tetrahedron numbered 1234 (of which there are, say, X) has one such face, each tetrahedron numbered 1123 or 1223 or 1233 (a total of, say, Y tetrahedra) has two such faces, and all the other tetrahedra have no such faces. Thus the total number of 123-faces is

$$X + 2Y.$$

On the other hand, every 123-triangle inside the large tetrahedron (say there are U such triangles) is a face of two small tetrahedra, one lying on each side of it. Suppose there are V 123-triangles lying on the face 123 of the large tetrahedron. Note that there are no 123-triangles on any of the other faces of the tetrahedron, by the conditions of the problem. It follows that

$$X + 2Y = 2U + V.$$

But the division of the face 123 of the large tetrahedron satisfies all the conditions of part **a**, and from the solution to that problem it follows that

V is odd. Hence X is also odd and therefore cannot be zero. Thus the theorem is proved.

Remark. This theorem can also be made more precise in the same way as the theorem of part a (see the note on the previous page). The formulation and proof of this refinement are left to the reader.

117. We shall prove more than is stated. We show that if an arbitrary polygon M is divided into triangles in accordance with the conditions of the problem (that no two triangles should adjoin along *part* of a side of one of them), and if an even number of triangles meet at each vertex of the decomposition D, then the vertices of D can be numbered 1, 2, or 3 in such a way that all the vertices lying on the boundary of M are numbered 1 or 2, and the three vertices of every triangle are numbered 123.

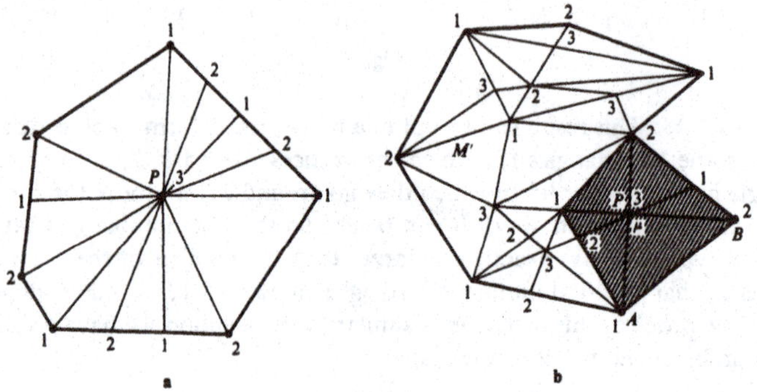

Fig. 58

The proof is by induction on the number of interior vertices.[7] Suppose first that there is only one vertex P inside M. By hypothesis, an even number of triangles meet at P, and therefore an even number of vertices lie on the boundary of M. Number the boundary vertices 1 and 2 alternately, and assign the number 3 to P (fig. 58).

Suppose next that there are n vertices inside M, and that the theorem holds for decompositions D' with fewer than n interior vertices. Let AB be one of the segments into which the vertices of D divide the boundary of M, and let ABP be one of the triangles of D. The vertex P must lie inside M; otherwise there would be only one triangle at one of the vertices A, B. By hypothesis, an even number of triangles meet at P: let us call the polygon composed of all these triangles μ (see fig. 58b, in which μ is shaded). Note that μ might be concave.

We now remove μ from M, thus obtaining a smaller polygon M', which clearly contains fewer interior vertices than M. Furthermore, the

[7] Since two or more triangles converge at each vertex of M, there must be at least one vertex inside M.

triangles into which M' is divided constitute a decomposition D' satisfying the conditions of the problem. For at every vertex of M' which is also a vertex of D, exactly two triangles have been removed; thus the number of triangles at each vertex of D' is even. By the induction hypothesis we can number all the vertices of D' so that the boundary vertices are all numbered 1 or 2 and so that every triangle is numbered 123. (We leave it to the reader to make the necessary changes in our argument if M' consists of a number of separate pieces.)

We now assign the number 3 to P and continue numbering the boundary of μ with 1's and 2's alternately (some of the boundary vertices will already have been numbered in this way). Since there are an even number of triangles, and therefore an even number of vertices, we can carry this out successfully. We now have an enumeration of the vertices of D, and by mathematical induction our theorem is proved.

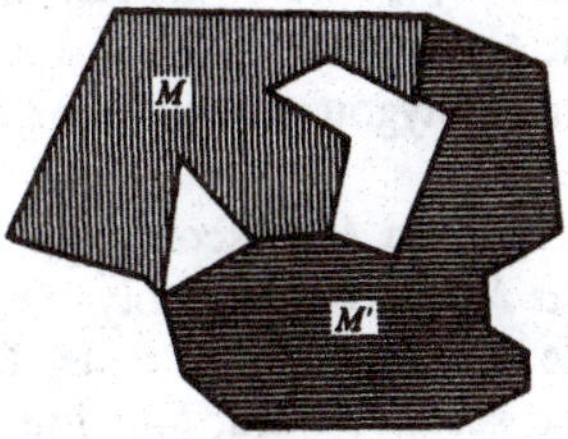

Fig. 59

118. It may be that there are cavities between two adjacent polygons M and M' of the decomposition D. Such cavities will, of course, be filled up by the other polygons of D (see fig. 59). In such a case the boundary between M and M' consists of a number of separate pieces. This phenomenon is inconvenient, and we proceed to construct a new decomposition of the square in which it does not take place. For this purpose we adjoin to M all the cavities between it and its neighbors; we obtain a larger polygon M_1. We then do the same with all the neighbors of M_1, and so on. Our new decomposition $\bar{D}$ is such that the boundary between any two pieces is either a single point or a connected polygonal line. It is clear that if some polygon of the new decomposition has at least six neighbors, then the same must have been true of the original decomposition, for our construction can only decrease the number of neighbors a polygon has. We note that each polygon $\bar{M}$ is contained within the polygon $\bar{\bar{M}}$ obtained by adjoining to M all its neighbors and all the gaps between it and its neighbors. Since by hypothesis both M and all its neighbors have diameter at most $\frac{1}{30}$, $\bar{\bar{M}}$ has a diameter of at most $\frac{3}{30} = \frac{1}{10}$ (fig. 60) and $\bar{M}$ has a diameter no greater.

Consider the polygon M_O of the new decomposition which covers the center O of the square (if O lies on a boundary, choose any polygon that

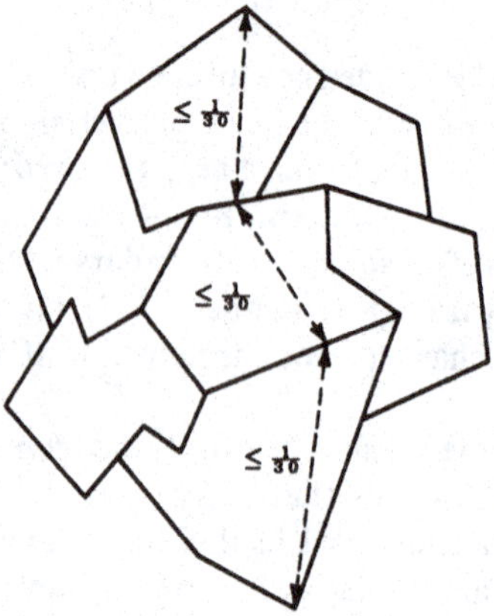

Fig. 60

covers it). We say that M_O has *rank 1.* All the polygons of the new decomposition adjacent to M_O we call *polygons of rank 2*; all other polygons adjacent to any polygon of rank 2 we call *polygons of rank 3*, and so on.

It is clear that the polygon M_O is contained inside a circle of radius $\frac{1}{10}$ and center at O; that M_O and all the polygons of the second rank are contained inside a circle of radius $\frac{2}{10}$ and center O, and, finally, that M_O and all polygons of the second, third, and fourth ranks are contained inside a circle of radius $\frac{4}{10}$ with center at O. It follows in particular that no polygon of the first four ranks can touch the boundary of the square.

We note some simple properties of the classification of polygons by rank.

(1) Every polygon of rank $n (n > 1)$ has at least one neighbor of rank $n - 1$.

(2) No polygon M of rank n has a neighbor of rank $<n - 1$ (otherwise M would have been assigned a rank $<n$). In other words, any two neighboring polygons have either the same rank or ranks differing by one.

(3) If the polygon M of rank $n > 1$ has less than two neighbors of rank n, then it has no neighbors of rank $n + 1$. For otherwise part of its boundary would be part of the boundary of a polygon of rank $n + 1$, while another part would be part of the boundary of a polygon of rank $n - 1$ [see property (1)]. These two polygons could not touch by property (2), and hence there would be at least two regions where M adjoined polygons of rank n, other ranks being impossible by property (2). But our decomposition is such that neighbors touch only along a single segment of a boundary, so that M must have at least two distinct neighbors of rank n.

We are now ready to prove that some polygon of $\bar{D}$ has at least six neighbors. The proof is by contradiction.

Suppose that every polygon of $\bar{D}$ has at most five neighbors. We will show that in that case no polygon of rank 4 can have neighbors of rank 5, or in other words, that there are no polygons of rank 5. This is a contradiction, for we have already seen that the square cannot be filled up by polygons of the first four ranks alone.

We consider separately two cases.

(a) If a polygon M of rank 4 has at most one neighbor of rank 4, then by property (3) it has no neighbors of rank 5.

(b) Suppose now the polygon M of rank 4 has at least two neighbors of rank 4. We show first that M has at least two neighbors of rank 3. For if not, and if M' is the single neighbor of rank 3, then M' has at least two neighbors of rank 4, apart from M: these two are the polygons M_1

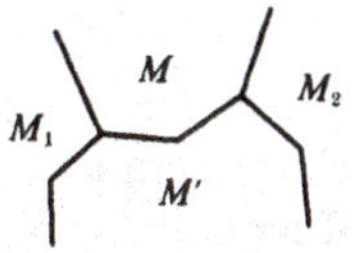

Fig. 61

and M_2 which border M at the ends of its common boundary with M' (fig. 61). M_1 and M_2 cannot be of rank 3, since by hypothesis M has no neighbors of rank 3 apart from M', and they cannot be of rank 5 by property (2). Moreover, M' has at least two neighbors of rank 3 [see property (3)] and at least one neighbor of rank 2 [see property (1)], a total of at least six neighbors, contrary to hypothesis.

We show next that M has at least *three* neighbors of rank 3. Suppose M' is any neighbor of M of rank 3 (we know there are at least two such). We show that M' has no neighbors of rank 4 apart from M. M' has at least two neighbors of rank 3 [see property (3)]. Moreover, it has at least two neighbors of rank 2. (This may be proved in exactly the same way as it was proved that M had at least two neighbors of rank 3.) And since the total number of neighbors of M is at most five by hypothesis, M must be the only neighbor of M' of rank 4. Now let M_1' and M_2' be the polygons which touch M' at the ends of the boundary it shares with M. These must be of rank 3 and are neighbors of M. So M has at least the three neighbors M', M_1', and M_2', or rank 3.

It is now clear that M cannot have neighbors of rank 5. For otherwise it would have at least two neighbors of rank 4 [by property (3)], and, as we have seen, three neighbors of rank 3, making a total of six neighbors, contrary to hypothesis. The theorem is proved.

Remark. It is easy to see that the square can be divided into arbitrarily small polygons, each of which has no more than six neighbors; the pattern of hexagons given in fig. 62 is an example.

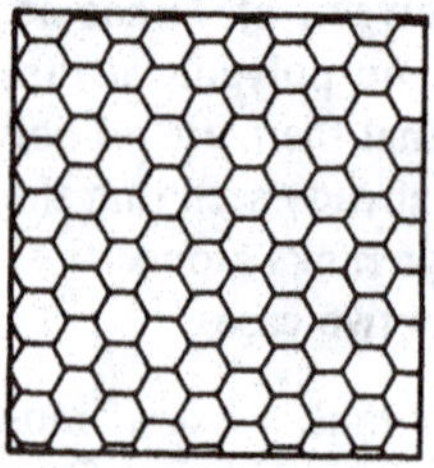

Fig. 62

IV. A PROPERTY OF THE RECIPROCALS OF INTEGERS

119. Let K be a continuous curve joining the points A and B. We will show that if K contains no chord parallel to AB of length a or length b, then it has no such chord of length $a + b$. It follows from this that K has a chord of length $1/n$. For otherwise it would have no chords of length

$$\frac{1}{n}+\frac{1}{n}=\frac{2}{n},\quad \frac{1}{n}+\frac{2}{n}=\frac{3}{n},\quad \frac{1}{n}+\frac{3}{n}=\frac{4}{n},\ldots$$

and, finally, it would have no chord of length

$$\frac{1}{n}+\frac{n-1}{n}=1,$$

whereas we know that it does contain such a chord (the chord AB itself).

We proceed to prove our assertion. The statement that K contains no chord of length a is equivalent to the statement that the curve K' obtained from K by a parallel displacement through a distance a in the direction AB has no point in common with K (fig. 63). Next, the curve K' has by hypothesis no chord of length b, for it is congruent to K. Thus the curve K'', obtained by translating K' a distance b in the direction AB, has no points in common with K'. We next show that the curves K and K'' do not intersect: this will mean that the curve AB has no chord of

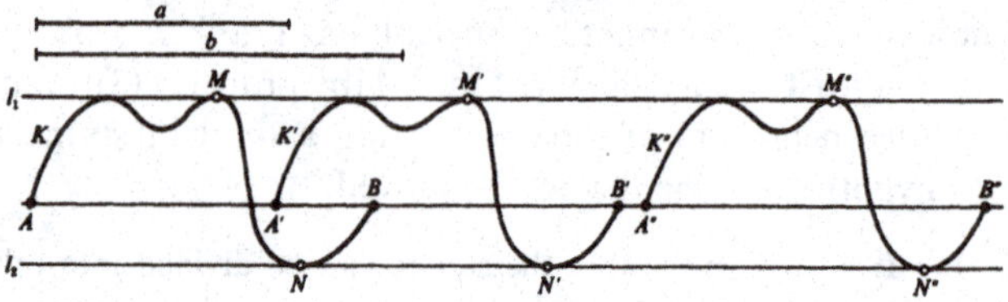

Fig. 63

length $a + b$ parallel to AB. (For K'' is obtained from K by translating it a distance $a + b$ in the direction AB.) The curve K' contains points lying on or above AB (for example, A and B themselves). Among these points there is at least one whose distance from AB is maximal; denote it by M' (fig. 63). Similarly, among the points of K' lying on or below AB, let N' be one whose distance from AB is maximal.

Through M' and N' draw lines l_1 and l_2 parallel to AB. (One or both of these lines may coincide with AB.) It is clear that all three curves K, K', K'' lie inside the strip enclosed by l_1 and l_2. The portion $M'N'$ of the curve K' divides this strip into at least two parts; and since K and K'' do not intersect K', each must lie entirely inside one of the parts. We claim that, in fact, they lie in *different* parts of the strip and therefore do not intersect.

For if M and M'' are the points of K and K'' corresponding to the point M' of K', then M and M'' lie on opposite sides of M'. Therefore, because they are continuous, the entire curves K and K'' lie on opposite sides of the curve $M'N'$ and so cannot intersect. This concludes the proof of the first half of the theorem.

We must now show that for each real a not of the form $1/n$, there is a continuous curve joining A to B and having no chord of length a parallel to AB. If $a > 1$ this is obvious: it suffices to require that the curve remains inside the region bounded by perpendiculars to AB erected at A and B. (See, for example, fig. 64.) If $1 > a > \frac{1}{2}$ we draw parallel lines through

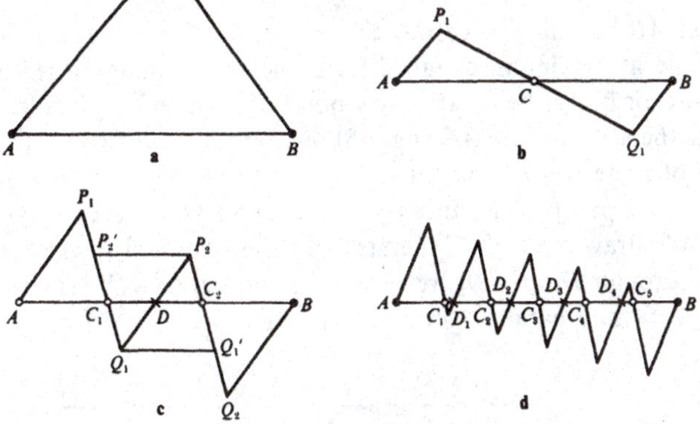

Fig. 64

A and B (neither of them the line AB itself), and an arbitrary line not parallel to these two through the midpoint C of AB. It is easy to check that the polygonal line AP_1CQ_1B (see fig. 64a) has no chord parallel to AB of length greater than $\frac{1}{2}$.

Suppose now that $\frac{1}{3} < a < \frac{1}{2}$. Divide AB into three equal parts $AC_1 = C_1C_2 = C_2B$ and also into two equal parts $AD = DB$. We draw three mutually parallel lines through A, D, and B, and through C_1 and C_2 two parallel lines not parallel to the first three. Then the polygonal line $AP_1C_1Q_1DP_2C_2Q_2B$ (fig. 64c) has no chords parallel to AB and of length lying between $\frac{1}{2}$ and $\frac{1}{3}$.

For if the ends of a chord lie on AP_1 and P_1Q_1, or on P_1Q_1 and Q_1P_2, or on Q_1P_2 and P_2Q_2, or on P_2Q_2 and Q_2B, then the chord is of length at most $\frac{1}{3}$ (since in the notation of fig. 64 $AC_1 = P_2'P_2 = Q_1Q_1' = C_2B = \frac{1}{3}$). If the ends of the chord lie on the segments AP_1 and Q_1P_2 or Q_1P_2 and Q_2B, then its length is $\frac{1}{2}$. Finally, in all other cases the length of a chord is greater than $\frac{1}{2}$.

In a completely analogous manner, we can construct a continuous curve having no chord parallel to AB and of length lying between $1/(n+1)$ and $1/n$. We divide AB into $n+1$ equal parts $AC_1 = C_1C_2 = \cdots = C_nB$ and independently into n equal parts $AD_1 = D_1D_2 = \cdots = D_{n-1}B$. We now draw arbitrary parallel lines through $A, D_1, D_2, \ldots, D_{n-1}, B$, and parallel lines intersecting these through $C_1, C_2, \ldots, C_n$. It is easy to see that the curve we obtain (see fig. 64d in which the case $n = 5$ is illustrated) satisfies the conditions: the proof is analogous to the one given for the case $n = 3$.

V. CONVEX POLYGONS

120a. Let AB be one side of the convex polygon M of area 1, and C a point of M at maximum distance from the line containing AB (C may be a vertex, or it may be an arbitrary point of a side of M parallel to AB). We draw the straight line AC (fig. 65) dividing M into two parts M_1 and M_2 (one of these will not exist if AC is a side of M). Suppose next that D_1 and D_2 are points of M at maximum distances from AC on either side of AC. We draw through C the straight line l parallel to AB, and lines l_1 and l_2 through D_1 and D_2, respectively, parallel to AC. The four lines AB, l, l_1, l_2 form a parallelogram P containing M.

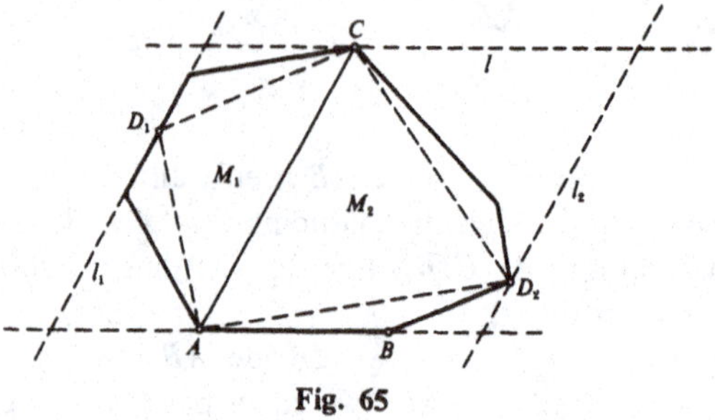

Fig. 65

Since M_1 and M_2 are convex, they contain the entire triangles AD_1C and AD_2C, respectively. The line AC divides P into two smaller parallelograms P_1 and P_2. It is clear that

$$\text{Area } (AD_1C) = \tfrac{1}{2} \text{ area } P_1 \quad \text{and} \quad \text{area } (AD_2C) = \tfrac{1}{2} \text{ area } P_2.$$

It follows from this that

$$\begin{aligned} \text{Area } P &= \text{area } P_1 + \text{area } P_2 = 2 \text{ area } (AD_1C) + 2 \text{ area } (AD_2C) \\ &\leqq 2 \text{ area } M_1 + 2 \text{ area } M_2 \\ &= 2 \text{ area } M = 2. \end{aligned}$$

If the area of P is less than 2, we can increase its area to 2 by translating one of the sides through a suitable distance. This larger parallelogram will still contain M.

b. Let P be a parallelogram containing the triangle ABC of area 1 (fig. 66). We can shrink this parallelogram by moving its sides together

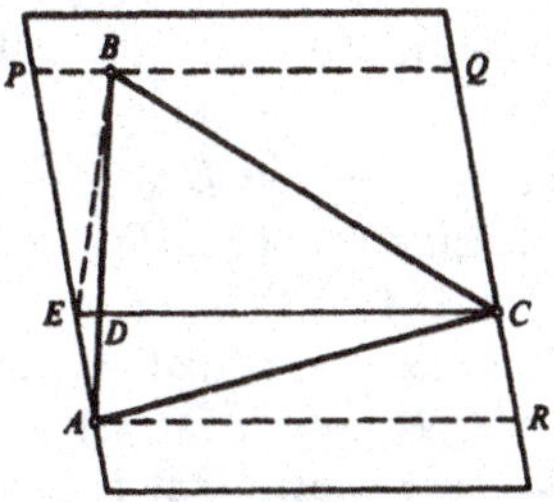

Fig. 66

parallel to themselves until each of them passes through a vertex of the triangle. Let A be the vertex (or one of the vertices) through which two sides of the parallelogram produced in this manner pass. Let the parallelogram be $APQR$, and let B and C lie on PQ and QR, respectively. Through C draw line CDE parallel to AR, meeting AB and AP in D and E, respectively. Denoting the area of any polygon M by $S(M)$, it is clear that

$$S(CBD) \leqq S(CBE) = \tfrac{1}{2}S(CQPE),$$

and

$$S(CAD) \leqq S(CAE) = \tfrac{1}{2}S(CRAE),$$

whence it follows that

$$\begin{aligned} S(ABC) &= S(CBD) + S(CAD) \\ &\leqq \tfrac{1}{2}S(CQRE) + \tfrac{1}{2}S(CRAE) = \tfrac{1}{2}S(APQR'); \end{aligned}$$

$$S(APQR) \geqq 2S(ABC) = 2.$$

(If C is an interior point of QR, then $S(APQR) = 2$ if and only if P and B coincide.)

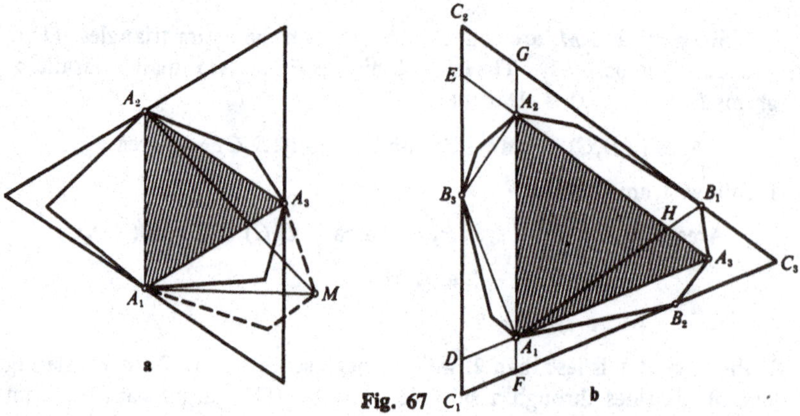

Fig. 67

121a. Inscribe in the given polygon U a triangle $A_1A_2A_3$ of maximum area[8] (fig. 67a). Through the vertices of the triangle draw lines parallel to the opposite sides. This gives a triangle T of area four times that of $A_1A_2A_3$. Suppose that the area of $A_1A_2A_3$ is $\leqq \frac{1}{2}$. Then the area of T is $\leqq 2$.

We now show that the polygon U lies entirely within T. Suppose some point M of U lies outside T. Then M is further from one of the sides (say A_1A_2) of the triangle $A_1A_2A_3$ than the opposite vertex (A_3). Hence the triangle A_1A_2M has a greater area than $A_1A_2A_3$ (see fig. 67a), contrary to the hypothesis that $A_1A_2A_3$ has the largest possible area for a triangle inscribed in U. Thus we have shown that U is enclosed within a triangle T of area not greater than 2, as required. (We could always expand T to obtain a triangle of area exactly 2, and it would also contain U.)

The case where the area of $A_1A_2A_3$ is greater than $\frac{1}{2}$ is slightly more complicated (fig. 67b). In each of the portions of U cut off by the sides of $A_1A_2A_3$, construct a triangle of maximum area with base a side of $A_1A_2A_3$. Through the other vertices B_1, B_2, B_3 of these three triangles, draw lines parallel to their bases. We obtain a larger triangle $C_1C_2C_3$. In exactly the same way as before we see that U lies within $C = C_1C_2C_3$.

We will show that $S(C_1C_2C_3) \leqq 2S(A_1B_3A_2B_1A_3B_2)$. Since

$$S(A_1B_3A_2B_1A_3B_2) \leqq S(U) = 1,$$

the required inequality then follows. Now $C_1C_2C_3$ is similar to $A_1A_2A_3$, so we can compute its area if we know the ratio C_1C_2/A_1A_2. To calculate this ratio, put

$$\frac{S(A_1A_2B_3)}{S(A_1A_2A_3)} = \lambda_3, \quad \frac{S(A_1A_3B_2)}{S(A_1A_2A_3)} = \lambda_2, \quad \frac{S(A_2A_3B_1)}{S(A_1A_2A_3)} = \lambda_1.$$

[8] It can be shown that such a triangle is one of the triangles whose vertices are vertices of U. See, for example, Shklyarskii *et al.* Ref. [21], chap. II, problem 23.

Note that

$$\lambda_1 + \lambda_2 + \lambda_3 = \frac{S(A_1A_2B_3) + S(A_1A_3B_2) + S(A_2A_3B_1)}{S(A_1A_2A_3)}$$

$$\leqq \frac{S(U) - S(A_1A_2A_3)}{S(A_1A_2A_3)} = \frac{1 - S(A_1A_2A_3)}{S(A_1A_2A_3)} < 1,$$

since by hypothesis $S(A_1A_2A_3) > \frac{1}{2}$. Hence

$$\frac{S(C)}{S(A)} = (\lambda_1 + \lambda_2 + \lambda_3 + 1)^2,$$

and since also it is clear that

$$\frac{S(A_1B_3A_2B_1A_3B_2)}{S(A)} = \frac{S(A) + S(B_1A_2A_3) + S(B_2A_1A_3) + S(B_3A_1A_2)}{S(A)}$$

$$= 1 + \lambda_1 + \lambda_2 + \lambda_3,$$

we finally arrive at the result

$$\frac{S(C)}{S(A_1B_3A_2B_1A_3B_2)} = 1 + \lambda_1 + \lambda_2 + \lambda_3 < 2,$$

since $\lambda_1 + \lambda_2 + \lambda_3 < 1$. As explained above, this completes the proof.

b. We first show that a square of side 1 cannot be enclosed in a triangle of area less than 2. Let $C = C_1C_2C_3$ be a triangle containing the square $A = A_1A_2A_3A_4$. We may suppose that at least three vertices of A lie on sides (or at vertices) of C, for otherwise we can simply shrink C appropriately until this occurs. Consider now the case where three of the vertices A_1, A_2, and A_3 lie on the sides of C opposite C_1, C_2, and C_3, respectively. We produce the sides A_4A_1 and A_4A_2 to meet C_3C_1 and C_3C_2 in F and D, respectively, and to meet C_1C_2 in G and E, respectively. Since the angles A_2A_4F and A_1A_4D are right angles, angles A_4FC_3 and A_4DC_3 are obtuse. Next,

$$\angle A_2EA_3 + \angle A_1GA_3 = 180° - \angle EA_4G = 90°;$$

$$\angle A_2A_3E + \angle A_1A_3G = 90°.$$

It follows that either $\angle A_2A_3E \leqq \angle A_2EA_3$ or $\angle A_1A_3G \leqq \angle A_1GA_3$. For definiteness suppose that the first of these inequalities holds. In that case we have

$$A_2E \leqq A_2A_3 = A_2A_4 < A_2D,$$

and it follows from fig. 68a that

$$S(A_2DC_3) > S(A_2EC_1), \qquad S(EDC_2) < S(C).$$

We have thus arrived at a circumscribed triangle EDC_2 of area smaller than C, one of whose sides contains two vertices of A, and therefore the side joining them. If this was the case to start with for C and A, then the preceding argument can be omitted.

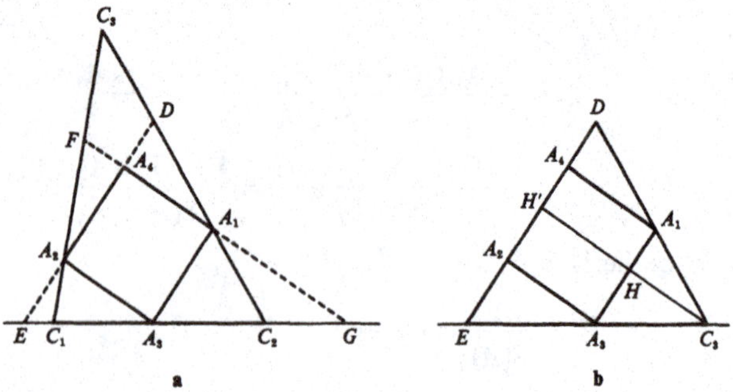

Fig. 68

Denote the height of triangle $A_1A_3C_2$ with base A_1A_3 by h, and let H and H' be the points of intersection of the perpendicular from C_2 with A_1A_3 and A_2A_4 (fig. 68b). Then $HH' = 1$ (the length of a side of the square) and

$$\frac{S(C_2HA_1)}{S(HA_1A_4H')} = \frac{h}{2}, \qquad \frac{S(C_2HA_1)}{S(A_1A_4D)} = h^2,$$

so that

$$\frac{S(A_1A_4D)}{S(HA_1A_4H')} = \frac{1}{2h}$$

and

$$\frac{S(C_2H'D)}{S(HA_1A_4H')} = \frac{S(C_2HA_1) + S(HA_1A_4H') + S(A_1A_4D)}{S(HA_1A_4H')} = 1 + \frac{h}{2} + \frac{1}{2h}.$$

In exactly the same way we may show that

$$\frac{A(C_2H'E)}{A(HA_2A_3H')} = 1 + \frac{h}{2} + \frac{1}{2h}.$$

It follows from this that

$$\frac{S(C_2ED)}{S(A)} = 1 + \frac{h}{2} + \frac{1}{2h} = 2 + \left(\frac{h}{2} - 1 + \frac{1}{2h}\right)$$

$$= 2 + \left(\sqrt{\frac{h}{2}} - \sqrt{\frac{1}{2h}}\right)^2 \geq 2,$$

that is, $S(C_2ED) \geq 2$, as was required.

We show now that if $A = A_1A_2A_3A_4$ is any rectangle of area 1, then a circumscribing triangle $C = C_1C_2C_3$ has area at least 2. This result follows easily from the corresponding result when A is a square. For given A and C in a plane P, let us project the whole configuration orthogonally onto a plane Q in such a way that the projection of A is a square (fig. 68a). If $A_1A_2 < A_2A_3$ we take Q to intersect P along A_1A_2, and to make an angle α with it, where $\cos \alpha = A_1A_2/A_2A_3$. The areas of $A' = A_1A_2A_3'A_4'$

and $C' = C_1'C_2'C_3'$ are equal to the areas of A and C multiplied by the factor $\cos \alpha$, for this is a property of the areas of all figures under orthogonal projection. If, then, $S(C) < 2\ S(A)$, we shall also have $S(C') < 2\ S(A')$, which is impossible, as we have seen.

From the result when A is a rectangle we can deduce the corresponding result when A is a parallelogram. Suppose the result is false for some parallelogram A and triangle C (fig. 69b). Construct a sphere whose diameter is equal to the longer diagonal A_1A_3 of A, and let A_2' and A_4' be the points of intersection of this sphere with perpendiculars erected at A_2 and A_4 to the plane π of A, and lying on opposite sides of this plane. Set $A' = A_1A_2'A_3A_4'$, and let C' be the triangle in the plane of A' whose

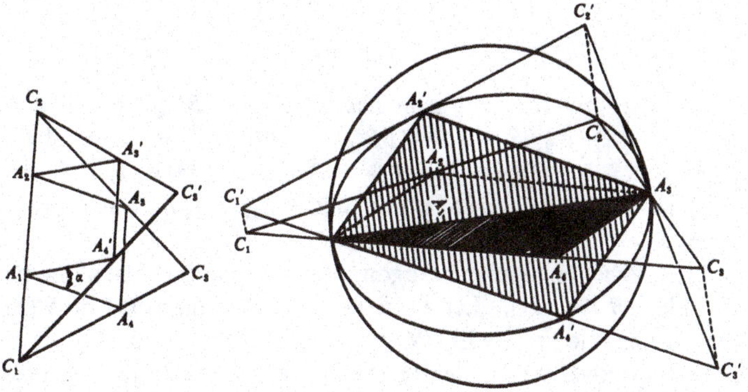

Fig. 69

orthogonal projection onto π is C. It is clear that A' is a rectangle whose orthogonal projection onto π is simply A, and we complete the argument in the same way as before.

122a. Draw two lines parallel to l lying on either side of the convex polygon M, and move them in until they pass through extreme points A and B of M. It is possible that the two lines l_1 and l_2 meet M not merely in a point but in a whole side (fig. 70). M now lies in the strip between l_1 and l_2; denote the width of this strip by d, and draw lines l_1', l_0, l_2' so as to divide it into four smaller strips of width $d/4$. Suppose the boundary of M meets l_1' in P and Q, and l_2' in R and S. (Note that these lines cannot intersect the boundary of M along a whole side of M because M is convex). Let p be the side of M passing through P (or one of the two such sides if P is a vertex), and, similarly, for q, r, and s. The area of the trapezoid T_1 bounded by the lines l_0, l_1, p, and q is $PQ \cdot d/2$; similarly, the area of T_2 bounded by l_0, l_2, r, and s is $RS \cdot d/2$. Since T_1 and T_2 together contain the whole of M,

$$S(M) \leq S(T_1) + S(T_2) = PQ \cdot \frac{d}{2} + RS \cdot \frac{d}{2} = (PQ + RS)\frac{d}{2}.$$

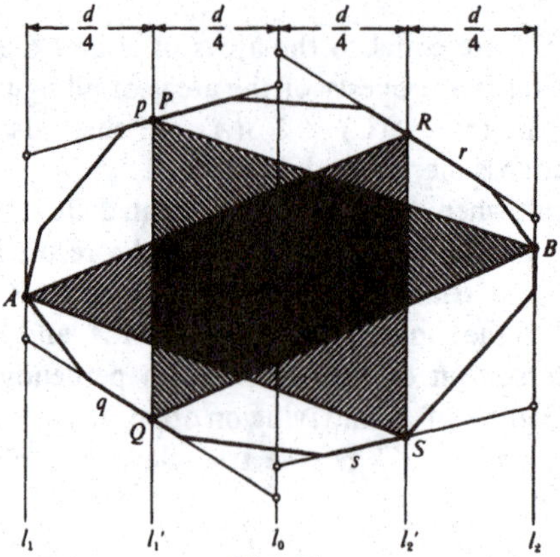

Fig. 70

Consider now the two triangles ARS and BPQ inscribed in M. Clearly $S(ARS) = \frac{1}{2}RS \cdot \frac{3}{4}d$; $S(BPQ) = \frac{1}{2}PQ \cdot \frac{3}{4}d$, so that $S(ARS) + S(BPQ) = (RS + PQ)\frac{3}{8}d = \frac{3}{4}(PQ + RS)\,d/2 \geq \frac{3}{4}S(M)$. It follows from this that either $S(ARS) \geq \frac{3}{8}S(M)$ or $S(BPQ) \geq \frac{3}{8}S(M)$, which proves the theorem.

b. Let M be the regular hexagon $ABCDEF$, and let l be a line parallel to the side AB (fig. 71). Let PQR be a triangle inscribed in M and of maximum possible area, with PQ parallel to AB. If P and Q lie on FA and BC, respectively, then clearly R must lie on DE (in fact, it makes no

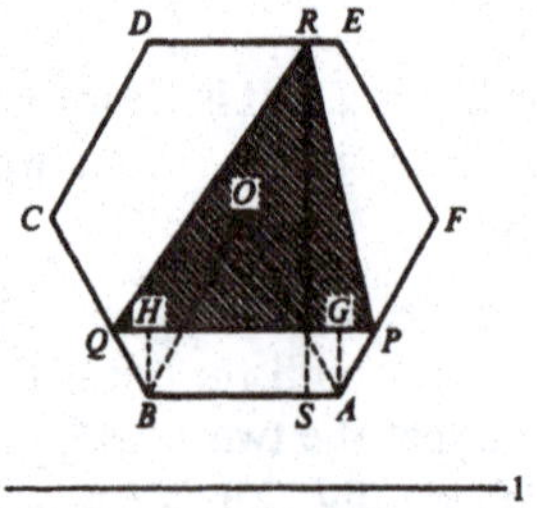

Fig. 71

difference where it lies on DE). Let us take the length of a side of the hexagon as our unit and write $AP = BQ = a$. Then (see fig. 71) we easily see that

$$PQ = AB + PG + QH = 1 + \frac{a}{2} + \frac{a}{2} = 1 + a,$$

and

$$h(PQR) = RS - AG = \sqrt{3} - \frac{a\sqrt{3}}{2} = (2 - a)\frac{\sqrt{3}}{2}.$$

Hence

$$S(PQR) = \frac{1}{2}(1 + a)(2 - a)\frac{\sqrt{3}}{2} = \frac{\sqrt{3}}{4}(2 + a - a^2)$$

$$= \frac{\sqrt{3}}{4}\left[2 + \frac{1}{4} - \left(a - \frac{1}{2}\right)^2\right].$$

It follows at once from this formula that the triangle PQR has maximum area when $a = \frac{1}{2}$, and then

$$S(PQR) = \frac{9}{4} \cdot \frac{\sqrt{3}}{4} = \frac{9\sqrt{3}}{16}.$$

But the area of the hexagon is given by

$$6S(OAB) = 6\frac{\sqrt{3}}{4} = \frac{3\sqrt{3}}{2}.$$

(O is its center.) It follows that the largest area a triangle inscribed in M and with one side parallel to AB can have is $\frac{3}{8}$ of the area of M.

VI. SOME PROPERTIES OF SEQUENCES OF INTEGERS

123a. We are given n arithmetic progressions, whose terms are integers. Supposing that every pair of progressions has a term in common, we must show that all n progressions have a term in common. We shall give a proof by induction. For $n = 2$ the theorem is obvious.

Suppose the theorem has already been proved for $n - 1$ progressions, and consider a set of n progressions, each pair of which has a common term. By the inductive hypothesis, the first $n - 1$ of them have a common term A. Now subtract A from all the terms of all the progressions: we obtain n new progressions, the first $n - 1$ of which have the term 0 in common. It is clear that if we can prove the theorem for these n new progressions, then it is also true for the original ones.

Let us denote the common differences of the n progressions by $d_1, d_2, \ldots, d_n$, where all the d_i are integers. Then the first $n - 1$ progressions consist of the multiples of $d_1, d_2, \ldots, d_{n-1}$, respectively, since they each contain the term 0. The general term of the nth progression is $a + kd_n$, where a is any one term and k is an integer which may be positive, negative, or zero.

We must show that there is a term $a + kd_n$ which belongs to all the other sequences, that is, that there exists an integer k such that $a + kd_n$ is divisible by all of the numbers $d_1, d_2, \ldots, d_{n-1}$. Equivalently, we must show that there exists an integer k such that $a + kd_n$ is divisible by the least common multiple N of the numbers $d_1, d_2, \ldots, d_{n-1}$.

Denote by D the greatest common divisor of the numbers N and d_n. Then there exist integers p and q (positive, negative, or zero) such that $D = pN + qd_n$. We leave the proof of this assertion to the end of the solution; meanwhile we assume its truth.

We prove now that a is divisible by D. Write D as a product of primes

$$D = p_1^{\alpha_1} p_2^{\alpha_2} \cdots p_r^{\alpha_r}.$$

The fact that D is divisible by $p_i^{\alpha_i}$ means that both d_n and N are divisible by $p_i^{\alpha_i}$. The fact that the least common multiple N of the numbers $d_1, d_2, \ldots, d_{n-1}$ is divisible by $p_i^{\alpha_i}$ means that some one of these numbers, say d_1, is divisible by $p_i^{\alpha_i}$. But, by hypothesis, the first and nth progressions have a term in common. In other words, there exist integers k' and k'' such that

$$k'd_1 = a + k''d_n,$$

or

$$a = k'd_1 - k''d_n,$$

from which it follows that a is divisible by $p_i^{\alpha_i}$ (since both d_1 and d_n are divisible by $p_i^{\alpha_i}$). Since this is true for each i, a is divisible by the product $D = p_1^{\alpha_1} p_2^{\alpha_2} \cdots p_r^{\alpha_r}$.

Multiply the equation $D = pN + qd_n$ by $m = a/D$, we obtain

$$a = pmN + qmd_n \quad \text{or} \quad a - qmd_n = pmN,$$

from which it follows that the term $a - qmd_n$ of the nth progression is divisible by N, which is what we had to show.

It remains for us to prove the existence of integers p and q such that $D = px + qy$, where D is the greatest common divisor of x and y. Consider the set I of all integers of the form $px + qy$, where p and q are any integers—positive, negative, or zero. It is easy to check that the sum or difference of any two members of I is again a member of I. [For $(p_1x + q_1y) + (p_2x + q_2y) = p_3x + q_3y$, where $p_3 = p_1 + p_2$; $q_3 = q_1 + q_2$.] It is also obvious that any multiple of a member of I is also a member of I. Now I has at least one positive member (for example, $x = 1 \cdot x + 0 \cdot y$). Let E be the smallest positive member of I. We intend to show that $E = D$. We first show that I consists precisely of all the multiples (positive, negative, or zero) of E. Suppose t is an arbitrary member of I. On dividing t by E we obtain a remainder r lying between 0 and $E - 1$.

We may write

$$t = nE + r, \qquad 0 \leqq r < E.$$

Now E is in I, so that nE is also, and t is in I, so that $r = t - nE$ is also in I, by the remark above. But we chose E to be the smallest positive member of I, and r is smaller. This can only be true if $r = 0$. Thus every member t of I is of the form nE. In particular, E is a common factor of x and y, both of which lie in I, and so divides D. On the other hand, every number of the form $px + qy$ is clearly divisible by D, and E is of this form (being a member of I). Since the two positive numbers D and E divide each other, they must be equal. We can therefore write

$$D = E = px + qy,$$

and equation (1) is proved.

Consider the two arithmetic progressions

$$P_1: \ldots, -2\sqrt{2}, -\sqrt{2}, 0, \sqrt{2}, 2\sqrt{2}, \ldots$$

and

$$P_2: \ldots, -2, -1, 0, 1, 2, \ldots$$

These progressions have the common term 0, and no other common terms. For if $k\sqrt{2} = k'$, where k and k' are nonzero integers, we deduce that $k'/k = \sqrt{2}$ is rational, a contradiction. We now construct a third progression having a term in common with each of the first two, but having no term equal to zero. Consider, for example, the progression

$$P_3: (1 + \sqrt{2}) + k(1 - \sqrt{2}) \quad (k = \ldots, -2, -1, 0, 1, 2, \ldots).$$

When $k = -1$ we get the term $2\sqrt{2}$, which is also in P_1. When $k = 1$ we get the term 2, which is in P_2. But no term of P_3 is equal to zero, since if

$$(1 + \sqrt{2}) + k(1 - \sqrt{2}) = 0,$$

then

$$k = \frac{\sqrt{2}+1}{\sqrt{2}-1} = \frac{\sqrt{2}+1}{\sqrt{2}-1} \cdot \frac{\sqrt{2}+1}{\sqrt{2}+1} = 3 + 2\sqrt{2}$$

$$= 5.82\ldots, \text{ a contradiction.}$$

Remark. It is easy to see that the first part of the problem remains true for arithmetic progressions of rationals (not necessarily integers). For one can find a number N such that when the progressions $a_i + kd_i$ $(i = 1, \ldots, n)$ are multiplied by N, they all become sequences of integers. We can take N, for example, to be the least common denominator of $a_1, a_2, \ldots, a_n, d_1, d_2, \ldots, d_n$. Neither the hypothesis nor the conclusion of the theorem is affected by multiplying or dividing all the progressions by the same number.

b. Suppose we are given n arithmetic progressions $P_1, \ldots, P_n$ of real numbers, every three of which have a term in common. Consider any two of the progressions, say P_1 and P_2. If they have only one term in common, then every other progression must also contain this term (for every other progression has a term in common with P_1 and P_2). If P_1 and P_2 have two or more terms in common, the difference between these terms can be written in either of the forms $k'd_1$ or $k''d_2$, where k' and k'' are integers and d_1, d_2 are the common differences of P_1, P_2. Thus

$$k'\,d_1 = k''\,d_2, \qquad \frac{d_1}{d_2} = \frac{k''}{k'},$$

so that d_1 and d_2 are commensurable (that is, have a rational ratio).

Therefore if the common differences of any two of the progressions are incommensurable, they have only one term in common, and this term also belongs to all the other progressions. On the other hand, if the common differences of *all* the progressions are commensurable, we can multiply $P_1, \ldots, P_n$ by a number so that the common differences of the resulting progressions $P_1', \ldots, P_n'$ are integers. We now subtract from all the terms of $P_1', \ldots, P_n'$ one of the terms of P_1', obtaining progressions $P_1'', \ldots, P_n''$. The entire progression P_1'' consists of integers (since the common difference is an integer and one of the terms is 0). Each of the other progressions $P_2'', \ldots, P_n''$ also contains an integer (since it has a term in common with the first) and therefore consists entirely of integers. Moreover, every pair of the progressions $P_1'', \ldots, P_n''$ has a term in common (since every three of them do), and therefore by the first part of the problem they all have a term in common. But this means that the original progressions also had a term in common, which is what we wanted to prove.

124a. The first part is nearly trivial. By symmetry, there is no loss of generality in assuming the first digit to be a 1. Then the second digit cannot be a 1, or 1 would be repeated, and therefore must be a 2. Our sequence thus starts with 12. The third digit must be 1 (since we cannot have two 2's in a row), and, similarly, the fourth must be 2. But then the sequence starts with 1212, and the subsequence 12 is repeated.

b. We show that each of the sequences I_n ($n = 1, 2, \ldots$), constructed by using the digits 1 and 2 in the manner explained in the hint at the end of the book, contains no digit or block of digits three times in succession. That I_n contains no three consecutive digits follows from the fact that it consists of a succession of 12's and 21's. In what follows we call the pairs 12 and 21 *links* of the sequences. It is harder to show that no block of digits in I_n can occur three times in a row. Our proof will be by the *method of descent:* we shall show that if I_n contains a block of digits repeated three times, then so does I_{n-1}. We thus see that $I_n, I_{n-1}, \ldots, I_3, I_2$ and finally I_1 contain such blocks, and this is obviously false, for $I_1 = 12$.

Case 1. The block P, which is repeated three times, contains an even number of digits. There are two subcases:

(i) The first time it appears, P starts at the first digit of a link. In this case it will start at the first digit of a link on the second and third appearances also. This means that P consists of an integral number of links. On replacing each link 12 by a 1 and each link 21 by a 2, P is replaced by a block Q half as long, which occurs three times in succession in I_{n-1}. Thus I_{n-1} contains a sequence Q of digits repeated three times, as required.

(ii) At its first appearance P starts with the second digit of a link (and therefore also on the second and third appearances). Suppose for definiteness that the first digit of P is a 1 (the argument is unaffected if 1 and 2 are interchanged throughout). In this case P starts with the second digit of the link 21 at each appearance, and therefore the last digit of P is a 2. We arrive at the configuration 1. It is clear from the configuration that

$$2\,\overbrace{1 * * * * \cdots 2}^{P}\,\overbrace{1 * * * * \cdots 2}^{P}\,\overbrace{1 * * * * \cdots 2}^{P}\,1$$

$$\underbrace{2\,1 * * * * \cdots}_{Q}\,\underbrace{2\,1 * * * * \cdots}_{Q}\,\underbrace{2\,1 * * * * \cdots}_{Q}\,2\,1$$

Configuration 1

I_n contains three successive blocks Q obtained by throwing away the final 2 of P and including an initial 2. But Q starts with the first digit of a link, so that case (ii) reduces to case (i).

Case 2. The block P contains an odd number of digits. Since I_n consists entirely of two-digit links, P contains an integral number of links plus one digit, and so must either begin or end in the middle of a link (but not both). If P begins in the middle of a link at its first appearance, then it begins at the beginning of a link at the second appearance, and therefore in the middle of a link at thc third appearance; if it begins at the beginning of a link at the first appearance, then it begins in the middle of a link at the second appearance, and at the beginning of a link at the third appearance. In either case we can find two successive appearances of P, the first beginning at the start of a link and the second in the middle. Call these two blocks P_1 and P_2.

Without loss of generality, we may suppose that P_1 starts with the link 12; otherwise we can merely interchange the roles of 1 and 2 in the argument. Since P_2 starts with the second digit of a link, and this digit is a 1, the link must be 21 and the last digit of P_1 is a 2 (see Configuration 2). Since the second digit of P_1 is a 2, so is the second digit of P_2, and this

$$\overbrace{1\,2 * * * * \cdots 2}^{P_1}\,\overbrace{1 * * * * \cdots 2}^{P_2}$$

$$1\,2\,1\,2 \qquad\qquad 2\,1\,2\,1$$

Configuration 2

means that the first link entirely inside P_2 is 21. Thus we see that the third digit of P_2, and therefore also of P_1, is a 1. This means that the second link of P_1 is a 12. We now see that the fourth digit of P_1, and therefore also of P_2, is a 2; this means that the second complete link of P_2 is a 21. Continuing in this way, we see that the links of P are all 12; this means that the last digit of P_1 must be a 1, for P contains an odd number of terms, and all the odd positions are occupied by 1's. This is a contradiction, and hence this case cannot occur. (We note, incidentally, that if P contained as many as six terms we would already have three successive links 12 in P_1.)

Remark. On writing out the sequences $I_1, I_2, I_3, \ldots$ it is easy to see that each sequence is an initial segment of the following sequence. Using mathematical induction it can be shown that in general the first 2^{n-1} digits of I_n constitute exactly I_{n-1}. It follows from this that we can not only write down arbitrarily long sequences of 1's and 2's in which no digit or block of digits occurs three times successively, but can actually write down an *infinite* sequence with this property.

125a. We construct sequences J_n ($n = 0, 1, \ldots$) in the manner described in the hints for this problem. The proof that none of these sequences contains a digit or block of digits twice in succession is analogous to that of problem 124b. First of all, we show that no sequence J_n can contain two successive equal digits. For J_n consists of links 02, 0121, 0131 and 03. Neither within one link, nor inside the sequence obtained by putting two links together, can two successive equal digits occur. For each link starts with a zero, whereas no link ends with a zero.

It remains to show that J_n contains no block of digits occurring twice in succession. The proof is by induction and follows closely the method used for problem 124b. It is clear that the sequence $J_0 = 01$ contains no repeated digit or block of digits. Suppose now that the sequence J_{n-1} also satisfies this condition, and that the sequence J_n *does* contain two successive equal blocks of digits. We show that these assumptions lead to a contradiction. Let J_n contain two successive blocks P, which we shall denote by P_1 and P_2. We now consider several cases.

Case 1. P_1 consists of an integral number of links. In this case it is clear that P_2 also consists of an integral number of links. We now replace every link of J_n by the corresponding number, or, in other words, go back to J_{n-1}. Then P_1 and P_2 become equal consecutive blocks of digits in J_{n-1}, a contradiction, since we supposed that J_{n-1} contained no two such successive blocks.

Case 2. The blocks P_1 and P_2 start with a digit a which in both cases occupies the same place in the same link. Suppose for definiteness that both blocks start with the 2 of the link 0121. In this case P_1 ends with the digits 01, and therefore so does P_2. We thus obtain configuration

3, from which we see that J_n also contains two successive equal blocks Q_1 and Q_2, each consisting of a number of complete links. We have already examined this case and shown that it is impossible.

$$P_1 \qquad\qquad P_2$$

$$0\,1\,2\,1\,*\,*\,*\,*\,0\,1\,2\,1\,*\,*\,*\,*\,0\,1$$

$$Q_1 \qquad\qquad Q_2$$

Configuration 3

Case 3. The blocks P_1 and P_2 start with the same digit a, which either appears in different links in P_1 and P_2, or in different positions of the same link. The digit a must be followed by the same digit b in both P_1 and P_2. Examining all possible cases (here it is important to know the digits that follow every given digit of a link), and bearing in mind that the last digit of a link is always followed by a zero, we see that this case can only occur when $a = 1$, and in one of the P's it is the 1 standing at the end of the link 0121, whereas in the other it is the 1 at the end of the link 0131. Suppose for definiteness that P_1 starts with the last digit of 0121, whereas P_2 starts with the last digit of 0131. (This is no loss of generality, for we may, if necessary, reverse the roles of 2 and 3 in the following argument.) The block P_1 ends in 013, the first three digits of the link 0131. Therefore P_2 also ends with 013. It is easily seen that the digits 013 can only appear in a sequence J_n if they are the first three digits of the link 0131. We conclude that the first digit after the end of P_2 is a 1, and we obtain configuration 4. We see from it that the sequence J_n contains two

$$P_1 \qquad\qquad P_2$$

$$0\,1\,2\,1\,0\,*\,*\,*\,*\,0\,1\,3\,1\,0\,*\,*\,*\,*\,0\,1\,3\,1$$

$$Q_1 \qquad\qquad Q_2$$

Configuration 4

successive equal blocks Q_1 and Q_2, lying one digit to the right of P_1 and P_2, and that these blocks consist of a number of complete links. Thus case (3) reduces to case (1).

Remark. In order to construct an *infinite* sequence from the digits 0, 1, 2, 3 satisfying the conditions of the problem (see the remark after the solution of **124b**), we consider instead of the sequence J_n the sequence J_n' obtained from it by the rule

$$J_0' = 01, \qquad J_n' = 01\tilde{J}_{n-1}.$$

(Here the tilde $\sim$ above a letter has the same significance as in the hint to this problem.) It is easy to check that the sequence J_{n-1}' constitutes the first part of the sequence J_n'.

b. The proof is similar to that of problems **124b** and **125a**, and like them, uses mathematical induction (the method of descent is really a type of inductive argument). It is clear that the sequence $K = 123$ contains no repeated digits or blocks of digits. Suppose now this is also true of the sequence K_{n-1} but false for the sequence K_n. We show that this leads to a contradiction. Suppose K_n contains two successive equal blocks P_1 and P_2. To begin with, it is clear that P_1 and P_2 cannot consist of a single digit. In fact, none of the individual links (that is, blocks 123, etc.) of K contains a repeated digit. Also, the last digit of one link cannot equal the first digit of the following link. Suppose, for example, that the links 231 and 123 stand next to each other. In that case, one of these links would occupy the place previously occupied by an odd-numbered term of K_{n-1}, and the other an even-numbered place, whereas they both have to occupy odd-numbered places. Next, suppose that the links 321 and 123 stand next to each other. Then they replace adjacent 1's in K_{n-1}, which is impossible by hypothesis. All other cases are argued in the same way as one or the other of these two.

We must now show that K_n cannot contain successive equal blocks P_1 and P_2. We consider separately a number of distinct cases and show that none of them can take place.

Case 1. The blocks P_1 and P_2 consist of complete links. In this case we replace all the links of K_n by the corresponding numbers, obtaining the sequence K_{n-1}, in which there will be two successive equal blocks (the replacements for P_1 and P_2), contrary to hypothesis.

Case 2. The number of digits in P_1 and P_2 is divisible by 3, but P_1 and P_2 do not consist of complete links. If P_1 starts with the second digit of a link, then so does P_2 (since the number of digits in P_1 is divisible by 3). If the two last digits of the link which starts P_1 are, say 12, then before them there stands a 3, and since P_2 must also start with the last two digits 12 of a link, the last digit of P_1 must be a 3. But then the last digit of P_2 is also a 3, and we obtain configuration 5. We see that the blocks Q_1

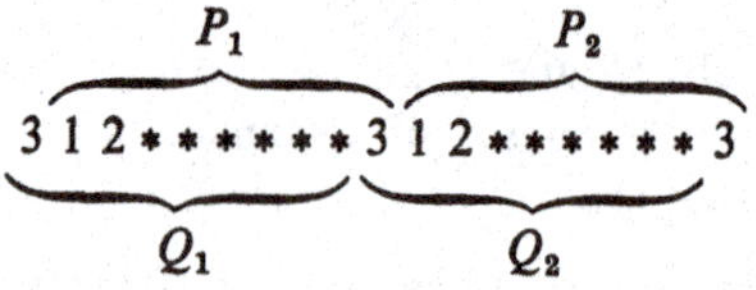

Configuration 5

and Q_2 lying one digit to the left of P_1 and P_2, respectively, are repeated adjacent blocks consisting entirely of complete links. Thus this case is reduced to the previous one. The case in which P_1 and P_2 start with the third digit of a link is dealt with in exactly the same way. The corresponding blocks Q_1 and Q_2 will in this case lie one digit to the right of P_1 and P_2.

Case 3. The number of digits in P_1 and P_2 is not a multiple of 3. Also, P_2 starts at the beginning of a link. We consider two subcases:

(i) The number of digits in the blocks P_1 and P_2 is equal to $3k + 1$. In this case P_1 and P_2 consist of an integral number of links plus one digit. The extra digits will be at the beginning of P_1 and at the end of P_2. Suppose for the sake of definiteness that the first digit of P_2 (and therefore also of P_1) is a 3. (Clearly the argument is unchanged if in fact the digit is a 1 or a 2, because of the symmetric role the three digits play in our construction.) The next two digits of P_2 will be a 2 and a 1, but we do not know in which order they occur. In the configuration 6 toward which we are working, the pair 12 or 21 (we do not know which) will be denoted by XX. It follows that the second and third digits of P_1 are also 12 or 21. But they are the first two digits of a link, and the third digit can only be a 3. Thus the fourth digit of P_1, and therefore also of P_2, is a 3. Now this digit is the first digit of a new link in P_2, the other members of which must be 21 or 12. We put them in P_2 and also in P_1. We can now argue as before that the next digit of P_1 is a 3, and so on. We conclude that the figure 3 stands in the 1st, 4th, 7th, . . . positions of P_1 and P_2, and since the total number of digits in P is $3k + 1$, both P_1 and P_2 must end with a 3. But then P_1 ends with a 3 while P_2 begins with one, so that we have two successive 3's, which is impossible, as we have shown.

$$\overbrace{3\,X\,X\,3\,X\,X\,3\cdots\cdots X\,X\,3}^{P_1}\;\overbrace{3\,X\,X\,3\,X\,X\,3\,X\,X\cdots\cdots 3}^{P_2}$$

Configuration 6

(ii) The number of digits in P_1 and P_2 is equal to $3k + 2$. In this case P_1 and P_2 consist of a number of complete links and two extra digits. These will stand at the beginning of P_1 and at the end of P_2. Suppose the first two digits of P_1 and P_2 are 1 and 2 in some order; then the third digit, being the third digit of a link, must be a 3 in P_2, and therefore also in P_1. But this means that the fourth and fifth digits of P_1 must be a 1 and a 2 (since they are the last two digits of a link beginning with a 3). And so the sixth digit of P_2 must be a 3, being the third digit of a link starting with 12 or 21, and so on.

We see (Configuration 7) that the blocks P_1 and P_2 have 3's in the

$$3\,\overbrace{X\,X\,3\,X\,X\,3\,X\,X\cdots\cdots 3\,X\,X}^{P_1}\;\overbrace{X\,X\,3\,X\,X\,3\,X\,X\,3\cdots\cdots X\,X}^{P_2}$$

Configuration 7

3rd, 6th, 9th, . . . places. It follows that P_1 ends with a link starting with a 3, that is, either 321 or 312. The block P_2 starts with a link whose third member is 3, that is, either 123 or 213. Thus the six middle digits of the block $P_1 \cup P_2$ must have one of the four forms 312213, 321123, 312123,

321213. All these are impossible, the first two because the middle two digits are the same, and the last two because the two links of which they are composed must both have come from odd-(respectively, even-) numbered digits in K_{n-1}.

The cases where the first two digits of P_1 are 1 and 3 or 2 and 3 are treated in the same way.

Case 4. The number of digits in P_1 and P_2 is not divisible by 3; P_2 does not begin at the beginning of a link. There are four subcases to consider.

(i) P_2 starts with the third digit of a link; P_1 and P_2 contain $3k + 1$ digits. In this case P_1 must start with the last two digits of a link and end with the first two digits of a link; and P_2 must start with the last digit of a link and end with a complete link.

Suppose for the sake of definiteness that P_1 and P_2 begin with a 3 (clearly this assumption involves no loss of generality). Then the last two digits of P_1 must be a 1 and a 2 (in some order), and therefore, the last two digits of P_2 must also be a 1 and a 2. Hence the third from last digit of P_2 (and so also of P_1) is a 3. But this means that the two digits before the 3 in P_1 must be a 2 and a 1, since the 3 is the last digit of a link. The argument continues in exactly the way it proceeded in cases 3(i) and (ii), and we see that a 3 stands in the 3rd, 6th, 9th, . . . positions from the end of P_1 and P_2 (Configuration 8). Since P_1 and P_2 contain $3k + 1$

$$\overbrace{3\,3\cdots\cdots X\,X\,3\,X\,X\,3\,X\,X}^{P_1}\overbrace{3\,3\cdots\cdots 3\,X\,X\,3\,X\,X\,3\,X\,X}^{P_2}$$

Configuration 8

digits, a 3 must stand in the second place of P_1 and P_2. But in P_1 we have a 3 in position 1, so that in this case we would have two 3's next to each other.

(ii) P_2 starts with the last digit of a link, and P_1 and P_2 contain $3k + 2$ digits. In this case P_1 must start with an entire link and P_2 must end on the first digit of a link. We suppose again, without loss of generality, that P_1 and P_2 start with a 3. Then the second and third places in P_1 (and therefore also in P_2) are occupied by a 1 and a 2 (in some order). But then the fourth position in P_2 (and therefore also in P_1) is occupied by a 3. Continuing the argument in exactly the same way as before, we conclude that the figure 3 stands in the 1st, 4th, . . . positions of P_1 and P_2 (Configuration 9). Since the number of digits is $3k + 2$, a 3 must stand in the next to last position in the two blocks, which contradicts the fact that the last two digits of P are a 1 and a 2 (since they are the first two digits of a link ending with the 3 at the beginning of P_2).

$$\overbrace{3\,X\,X\,3\,X\,X\,3\,X\,X\cdots\cdots 3\,X}^{P_1}\overbrace{3\,X\,X\,3\,X\,X\,3\,X\,X\,3\cdots\cdots 3\,X\,3}^{P_2}$$

Configuration 9

(iii) P_2 starts with the last two digits of a link; P_1 and P_2 contain $3k + 1$ digits. In this case P_1 must start at the beginning of a link and P_2 end with the first two digits of a link. Suppose for the sake of definiteness (and without loss of generality) that P_1 and P_2 end with a 3. Then the first two digits of P_2 (and P_1) are a 1 and a 2. Arguing as in the preceding cases, we find that a 3 stands in the 3rd, 6th, 9th, . . . positions of P_1 and P_2. Since the total number of digits in these two blocks is $3k + 1$, a 3 stands in the next to last place. Since there is also a 3 in the last place, we have two 3's next to each other, which is impossible.

$$\overbrace{X\,X\,3\,X\,X\,3\,X\,X\,3\cdots\cdots\;3\,3}^{P_1}\overbrace{X\,X\,3\,X\,X\,3\,X\,X\;\cdots\cdots\;3\,3}^{P_2}$$

Configuration 10

(iv) P_2 starts with the last two digits of a link, and P_1 and P_2 contain $3k + 2$ digits. In this case P_1 starts with the last digit of a link, and P_2 ends with a complete link. Suppose once again that P_1 and P_2 end with a 3. Then, arguing exactly as before, we find that P_1 and P_2 have a 3 in the 1st, 4th, 7th, . . . places counting from the end (Configuration 11).

$$\overbrace{X\,3\;\cdots\cdots\;3\,X\,X\,3\,X\,X\,3}^{P_1}\overbrace{X\,3\;\cdots\cdots\;X\,X\,3\,X\,X\,3}^{P_2}$$

Configuration 11

Since the number of digits in the blocks is $3k + 2$, it follows in particular that there is a 3 in the second position of P_2. But this position is occupied by the last digit of the link whose first digit is the final 3 in P_1, and this is a contradiction.

We have now examined all possible cases, and proved that none of them can occur. The proof is thus complete.

Remark. Suppose we start with the sequence $123\cdots n$, and construct further sequences by successively replacing the digits standing in odd-numbered places as follows:

1 by the block $123\cdots n$
2 by the block $234\cdots n1$
3 by the block $345\cdots n12$
. . . .
n by the block $n123\cdots(n-2)(n-1)$

and the digits standing in even-numbered places as follows:

1 by the block $n(n-1)(n-2)\cdots 321$
2 by the block $1n(n-1)(n-2)\cdots 432$
3 by the block $21n(n-1)(n-2)\cdots 543$
. . . .
n by the block $(n-1)(n-2)(n-3)\cdots 321n$.

We can show by a proof analogous to 125**b** above that none of the sequences (of digits 1, 2, 3, . . . , n) obtained in this way contains a repeated digit or block of digits.

126. We show that the number $T = T_n$ constructed according to the rule given in the hints does in fact satisfy the conditions of the problem. From the construction of T_n we know that no two n-digit numbers selected from it are identical. Thus to complete the solution it suffices to show that any sequence S of n 0's and 1's occurs somewhere in T. The proof is by induction on the number of 1's at the end of S.

It follows from the construction of T_n that it contains n consecutive 1's, for that is the way it starts. The only sequence ending with exactly $(n - 1)$ ones is $0\underbrace{11 \cdots 11}_{n-1 \text{ times}}$. We claim that this number is to be found in T_n, and in fact that it comprises the last n digits of T_n. For let us denote the last n digits of T_n by $\alpha_1\alpha_2\alpha_3 \cdots \alpha_n = \alpha_1\beta$, say, where β is the sequence $\alpha_2\alpha_3 \cdots \alpha_n$. By the definition of T_n the n-digit numbers $\beta 0$ and $\beta 1$ have already occurred as n successive digits of T_n. For otherwise we could write either a 1 or a 0 after β, and β would not lie at the end of T_n. Suppose now that not all the digits of β are 1's. Then neither of the sequences $\beta 0$ and $\beta 1$ can start T, for T starts with n 1's. It follows that the sequences $\beta 0$ and $\beta 1$ must both occur somewhere in the *interior* of T. Since no sequence of n digits can repeat itself in T, β must be preceded once by a 1 and once by a 0. But in that case α_1 cannot be either a 1 or a 0 (for in either case the final sequence, 1β or 0β, would already have occurred). This is a contradiction, and shows that β contains only 1's. Thus the last $n - 1$ digits of T are 1's, and the nth from the last digit must be a 0 (or the sequence 1β would occur twice: once at the beginning and once at the end of T).

We have thus shown that the sequence $011 \cdots 11$ does indeed occur in T.

Let us write $\alpha(r)$ for a sequence of r zeros and 1's (about whose distribution we know nothing), and $\beta(r)$ for a sequence of r ones. Let us suppose that every $\alpha(n)$ of the form $\alpha(n - i - 1)0\beta(i)$, where $i > m$ has already been shown to occur in T. We proceed to show that the number $k = \gamma\delta\, 0\beta(m)$ also occurs in to T, where γ is an $\alpha(1)$ and δ an $\alpha(n - m - 2)$.

By our inductive hypothesis $\delta 0\beta(m + 1)$ occurs somewhere in T. But by the construction of T this number can occur only if somewhere earlier in T there is the sequence $\delta 0\beta(m)0$. We see therefore that $\theta = \delta 0\beta(m)$ occurs twice in T, and neither time at the beginning, since it contains a zero. And since no $\alpha(n)$'s can occur twice in T, θ must be preceded on one occasion by a 1 and on the other occasion by a zero. Thus whatever the value of γ, $k = \gamma\theta$ occurs in T. The proof is thus complete.

Remark. Since the number of distinct $\alpha(n)$'s is clearly 2^n, it follows in particular from this solution that T_n contains $(2^n + n - 1)$ digits. For if it contained any fewer, it would be missing at least one $\alpha(n)$, whereas if it contained any more, at least one $\alpha(n)$ would occur more than once.

We note that one can use a similar construction to obtain a number $L = L_n$ in the decimal system, such that every possible sequence of n successive digits chosen from it is different, and such that every sequence of n digits is found somewhere in it. We start L_n by writing down n 9's, and thereafter write in each position the smallest digit which will not introduce duplicated sequences of n digits in the part of L already written down, until we can go no further.

VII. DISTRIBUTION OF OBJECTS

127. We prove a more general proposition: that if we are given n cookies of each of m different flavors, and if we put n cookies into each of m different boxes, then we can always take one cookie from each box so that no two of the cookies we take are of the same flavor (thus we have one cookie of each flavor). The assertion is clearly true when $m = 1$ (one box and one kind of cookie) and when $n = 1$ (one of each flavor and one cookie per box). We show now that the theorem is true for any value of n.

Consider first the case where $n = 2$ (two cookies per box). In this case it is easy to describe a process of selection satisfying the conditions of the problem. Select one of the cookies from the first box. Say this is a cookie of type 1, and say the other cookie in the first box is of type 2 (we shall deal with the case where this second cookie is also of type 1 shortly). Since we have two cookies of each kind, some box contains the second cookie of type 2. Say this is the second box, and the other cookie in it is of type 3. Then we take the cookie of type 2 from the second box, and find box 3, containing the second cookie of type 3. Say box 3 also contains a cookie of type 4: we take out the cookie of type 3 and look for a fourth box containing the second cookie of type 4. We continue in this way until we finally come to a box, say the kth, which contains the second cookie of type 1. If $k = 1$ we have the case (which we promised to deal with) where both cookies of the first type lie in the same box; if $k = m$ then we have finished our process of selection. In any case we have found k boxes which contain cookies of the first k kinds, and no others, and none of the untouched boxes contain any of these types of cookies. So we forget about the first k types of cookies and the first k boxes, and start the whole process again on the remaining boxes (see Configuration 12). On this second attempt we may still fail to get one of every type of cookie, but in that case we merely have to start once more.

Sooner or later we will obtain one of every type of cookie. (We could have presented this proof more elegantly in the form of an argument by induction on m: when we have finished the first cycle we have again the situation we started with but a smaller m.)

1	2	3		k		$k+1$	$k+2$	
$1 \mid 2$	$2 \mid 3$	$3 \mid 4$	$\cdots$	$k \mid 1$	|	$k+1 \mid k+2$	$k+2 \mid k+3$	$\cdots$

Configuration 12

For the cases with $n \geqq 3$ we shall give a proof by induction on n. Suppose that for all n smaller than some given value (greater than 2) we have already proved the theorem. Then we will show that the theorem holds for this value also.

We are given n cookies of each of m different types, $m \cdot n$ cookies in all. Suppose for the moment that the arrangement of the cookies in boxes is such that it *is* possible to make a selection in accordance with the conditions of the problem. We show that if we interchange any two cookies, the new arrangement of cookies in the boxes is still such that we can make an admissible selection.

For by assumption we can select m cookies, one of each kind, from the m boxes. After that we have an arrangement of $(n - 1)$ cookies of each of m kinds, arranged so that $(n - 1)$ cookies lie in each of the m boxes. By our inductive hypothesis we may make a second selection of m cookies, one of each kind, and one from each box. After this there will be $(n - 2)$ cookies of each of m kinds lying in m boxes, $(n - 2)$ to a box. (Remember that $n \geqq 3$.) By our inductive hypothesis we can make a *third* admissible selection. Suppose now that we change the initial arrangement by interchanging two of the cookies. In this process at least one of the three sets of m cookies that we could select will be undisturbed, so that every one of the m cookies in this selection remains in its original box. So we merely select this set.

We can now conclude that whatever the initial arrangement of the cookies in the boxes was, we can always make an admissible selection. For consider the arrangement in which each box is filled by all the cookies of some one sort. It is clear that we can make an admissible selection from it. (In fact we cannot *help* making an admissible selection if we take one cookie from each box.) Now *any* arrangement of the cookies in the boxes can be obtained from this arrangement by successive interchanges of pairs of cookies. Thus by the theorem we proved above, applied a sufficient number of times, we conclude that *any* arrangement permits an admissible selection. The general result now follows by mathematical induction.

Remark. This problem can also be stated in a geometrical form. Suppose we have two rows of m points (as in fig. 72, where $m = 4$).

Now suppose that lines are drawn, each of which connects a point of the

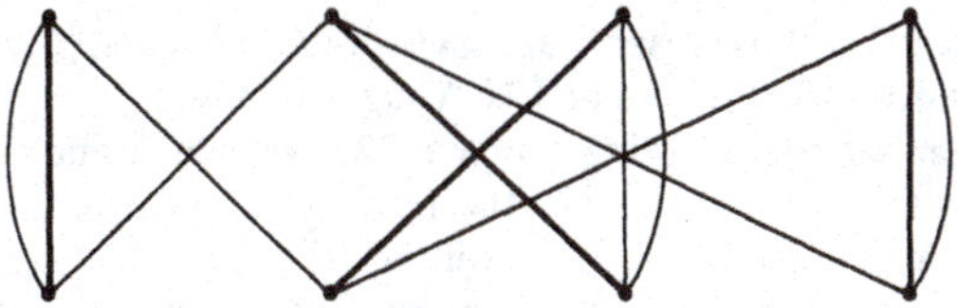

Fig. 72

top row with a point of the bottom row. It is permitted that the same pair of points be connected by more than one line. We require, however, that the same number of lines, say n, pass through every point (in fig. 72, $n = 3$). We assert that it is then possible to choose m of the lines such that every point lies on exactly one of them (in fig. 72 the heavy lines form such a choice).

To see that this result is equivalent to problem 127, think of the points $A_1, \ldots, A_m$ on the top row as the m flavors, and the points $B_1, \ldots, B_m$ on the bottom row as the m boxes. The lines are the cookies. Whenever a line connects A_i to B_j, we put a cookie of flavor A_i in the box B_j. The fact that n lines meet at each A_i means that there are n cookies of each flavor, and that the fact that n lines meet at each B_j means that there are n cookies in each box. A choice of m lines such that each point lies on one of them is the same as a choice of m cookies, one of each flavor and one in each box.

In its geometrical form, problem 127 is a famous theorem of graph theory, due to König. For other proofs, see Ref. [16].

128. We note that the hypothesis of the problem is clearly necessary in order to marry off the boys (for if it is not satisfied, then there is some set of k boys who are collectively acquainted with fewer than k girls). To prove sufficiency we use induction on m. For $m = 1$ the result is trivial. Suppose that $m > 1$, and that the theorem has been proved when the number of boys is $<m$. We distinguish two cases:

(1) Any k boys, where $k < m$, have at least $k + 1$ acquaintances.
(2) There is a set S of k boys, $k < m$, who have exactly k acquaintances.

In case (1) pick one of the boys and marry him to one of his acquaintances. There remain $m - 1$ boys, and any k of them have at least k acquaintances among the $M - 1$ remaining girls. By induction these $m - 1$ boys can be married off, and we are through.

In case (2) we can marry the boys of the set S by the induction hypothesis. There remain $m - k$ boys, and we assert that they satisfy the conditions of the problem with respect to the set R of remaining girls. To see this, consider any subset T containing l of these boys, where $1 \leqq l \leqq m - k$. If the boys in T were acquainted with fewer than l girls of R, then the boys in $S \cup T$ would be acquainted with fewer than $k + l$ girls, a contradiction to the hypothesis. Hence the boys in T are acquainted with at least l girls of R. By induction we can marry them off, and the proof is complete.

Remark. The above result was first proved by P. Hall. The proof we have given here is due to P. Halmos and H. Vaughan, (Ref. [8]).

As in the remark following problem 127, we can formulate the marriage problem geometrically. Represent the girls by a row of points $A_1, \ldots, A_M$, and the boys by points $B_1, \ldots, B_m$ as in fig. 73 (where $m = 4$, $M = 6$).

Connect A_i and B_j if the boy B_j is acquainted with the girl A_i. The hypothesis says that any k points on the bottom row are connected to at least k points on the top row. The conclusion is that m of the lines can be chosen so that each A_i is on at most one, and each B_j on exactly one of them. (In fig. 73, the heavy lines form such a choice.)

It is now easy to show that problem 127 is a special case of 128. Recall that in problem 127 we have $M = m$. Moreover, from each subset of k points on the bottom row there emanate exactly kn lines. These lines must terminate on at least k different points of the top row, since at most n of them can go through any given A_i. Thus the conditions of problem 127 imply those of 128. This gives us another solution to problem 127.

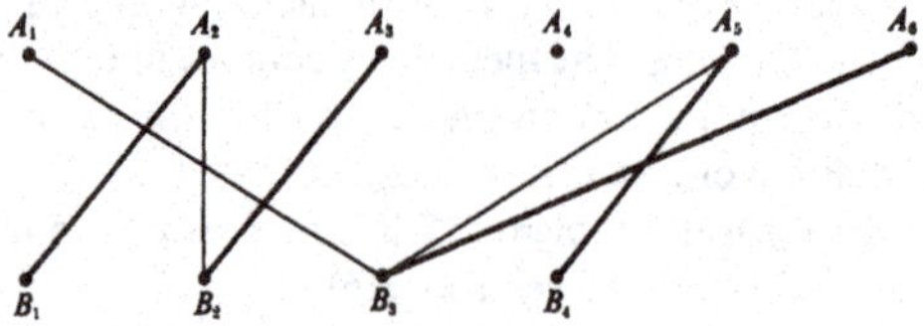

Fig. 73

VIII. NONDECIMAL COUNTING

129. We will show that the number in the $(A + 1)$st row and $(B + 1)$st column is the Nim sum $A \oplus B$ described in the hints at the back of the book. Before doing this, we will derive some properties of $A \oplus B$.

(1) The Nim sum is commutative, i.e. $A \oplus B = B \oplus A$. This follows at once from the definition.

(2) The Nim sum is associative, i.e. $(A \oplus B) \oplus C = A \oplus (B \oplus C)$. In fact if $A = \langle a_n a_{n-1} \cdots a_1 \rangle$, $B = \langle b_n b_{n-1} \cdots b_1 \rangle$, $C = \langle c_n c_{n-1} \cdots c_1 \rangle$, we easily see from the definition that both $(A \oplus B) \oplus C$ and $A \oplus (B \oplus C)$ are equal to $\langle d_n d_{n-1} \cdots d_1 \rangle$, where

$$d_i = \begin{cases} 0 & \text{if} \quad a_i + b_i + c_i \quad \text{is even} \\ 1 & \text{if} \quad a_i + b_i + c_i \quad \text{is odd} \end{cases}$$

(3) $0 \oplus A = A$ for all A, i.e. 0 is an *identity element* under $\oplus$.

(4) $A \oplus A = 0$ for all A (since if $A = \langle a_n a_{n-1} \cdots a_1 \rangle$, then $a_i + a_i = 2a_i$, which is even for all i). This means that each integer A is its own

inverse under $\oplus$. In the language of abstract algebra, properties (1)–(4) imply that the nonnegative integers form a *commutative group* under the Nim sum. In view of (2) we can simply write $A \oplus B \oplus C$ to denote either of the numbers $(A \oplus B) \oplus C$ or $A \oplus (B \oplus C)$.

We are now ready to prove that $A \oplus B$ is the entry in the $(A + 1)$st row and $(B + 1)$st column of the array under consideration. To do this we have to check that $A \oplus B$ has the following properties:

(a) $0 \oplus 0 = 0$.

(b) $A \oplus B$ is different from all the numbers $P \oplus B$ $(P < A)$ and $A \oplus Q$ $(Q < B)$.

(c) Any nonnegative integer $X < A \oplus B$ is equal to one of the numbers $P \oplus B$ $(P < A)$ or $(A \oplus Q)$ $(Q < B)$.

Property (a) is clear. We proceed to prove (b). If $A \oplus B = P \oplus B$, where $P < A$, then $A \oplus B \oplus B = P \oplus B \oplus B$ Since $B \oplus B = 0$, we have $A = P$, a contradiction. Similarly, the equation $A \oplus B = A \oplus Q$ where $Q < B$ leads to a contradiction.

To prove (c), let $A = \langle a_n a_{n-1} \cdots a_1 \rangle$, $B = \langle b_{n-1} \cdots b_1 \rangle$, $A \oplus B = C = \langle c_n c_{n-1} \cdots c_1 \rangle$, and suppose $X = \langle x_n x_{n-1} \cdots x_1 \rangle$ is any nonnegative integer $< C$. Let k be the greatest integer such that $c_k \neq x_k$; since $C > X$ we must have $c_k = 1$ and $x_k = 0$. Since $c_k = 1$ we have either $a_k = 1$, $b_k = 0$ or $a_k = 0$, $b_k = 1$. Suppose that $a_k = 1$, $b_k = 0$. Then $B \oplus X < B \oplus C$, since $b_k + x_k = 0 < b_k + c_k = 1$, while all "higher" digits of $B \oplus X$ and $B \oplus C$ are the same. But $B \oplus C = B \oplus B \oplus A = A$; thus, $B \oplus X < A$. Putting $P = B \oplus X$, we have $P \oplus B = B \oplus B \oplus X = X$, which shows that X is of the desired form $P \oplus B$ with $P < A$. Similarly the case $a_k = 0$, $b_k = 1$ leads to $X = A \oplus Q$ where $Q < B$.

This concludes the proof. It remains only to observe that in binary notation, $999 = \langle 1111100111 \rangle$ and $99 = \langle 1100011 \rangle$, and therefore the number at the intersection of the 1000'th row and the 100'th column is $999 \oplus 99 = \langle 1110000100 \rangle = 900$.

130. Suppose the three piles have a, b, c matches in them. We write the numbers to base 2 in the form

$$a = a_m 2^m + a_{m-1} 2^{m-1} + \cdots + a_1 2 + a_0$$

$$b = b_m 2^m + b_{m-1} 2^{m-1} + \cdots + b_1 2 + b_0$$

$$c = c_m 2^m + c_{m-1} 2^{m-1} + \cdots + c_1 2 + c_0,$$

where each of the digits $a_0, b_0, c_0, a_1, b_1, c_1, \ldots, a_m, b_m, c_m$ is either 0 or 1. We have given a, b, and c the same number of digits; this is possible provided we allow an initial digit to be zero. We require one of the initial digits to be a 1, but not necessarily all of them. A player on making a move changes just one of the numbers a, b, c to a smaller number. In the process he must change at least one of the digits of this number. Thus, if he takes some matches from the first pile, he must change at least one of the digits $a_0, a_1, \ldots, a_m$. Note now that every change in a digit reverses its parity (that is, changes it from even to odd or odd to even). It follows that a player taking a match from the first pile necessarily changes the parity of at least one of $a_0, a_1, \ldots, a_m$. A similar conclusion holds if he takes matches from the second or third piles. Consider now the sums

$$a_m + b_m + c_m, a_{m-1} + b_{m-1} + c_{m-1}, \ldots, a_1 + b_1 + c_1, a_0 + b_0 + c_0.$$

If at least one of these sums is odd, we assert that the first player can win, whatever his opponent does. Suppose, for example, that the first odd sum is $a_k + b_k + c_k$. Then at least one of the terms a_k, b_k, c_k is equal to 1; for the sake of definiteness let us suppose $a_k = 1$. Then by taking a suitable number of matches from the first pile the first player can make sure that none of the digits $a_m, a_{m-1}, \ldots, a_{k+1}$ changes, while the digits $a_{k-1}, \ldots, a_1, a_0$ assume any values he likes, and a_k changes from 1 to zero. This is so because any number satisfying all these conditions must be less than a, and therefore obtainable from a by the subtraction of a suitable number. In particular, the first player can make sure that all the sums

$$a_k + b_k + c_k, a_{k-1} + b_{k-1} + c_{k-1}, \ldots, a_1 + b_1 + c_1, a_0 + b_0 + c_0$$

become even (whether or not they were even before). When the second player makes his move, he must alter the parity of at least one term, and therefore also of at least one of the sums

$$a_m + b_m + c_m, a_{m-1} + b_{m-1} + c_{m-1}, \ldots, a_1 + b_1 + c_1, a_0 + b_0 + c_0.$$

This means that after his turn there will once again be an odd sum for the first player, and he now moves as before. As the game continues in this manner, the number of matches continually decreases. Since at the moment when all the matches have gone, the sums are all zero, and hence even, the first player must have made the last move.

If in the initial position all of the sums

$$a_m + b_m + c_m, a_{m-1} + b_{m-1} + c_{m-1}, \ldots, a_1 + b_1 + c_1, a_0 + b_0 + c_0$$

are even, then the first player cannot guarantee a win; in fact the second player can win by playing in the manner described above. Of course, the first player can move at random, waiting for the second player to make a mistake: after a single mistake on the second player's part the first player can win.

Remark 1. Using the Nim sum which was discussed in problem 129, the strategy can be described as follows. Compute the Nim sum $a \oplus b \oplus c$ of the three piles. If it is not zero, you can always move to a position where it *is* zero. Then, no matter what your opponent does, he will leave you in a position where the Nim sum is *not* zero. Continue in this way, always leaving your opponent in a position where the Nim sum is zero. Finally he will be left in the position $a = b = c = 0$, and then you will have won the game. This is the reason for calling $A \oplus B$ the "Nim sum" of A and B.

Remark 2. Note that the game has a bias in favor of the first player: losing positions for him are in a certain sense exceptional (since winning positions occur far more frequently, especially when the piles are large). Thus if two players who know how to take advantage of a winning position play each other, the first to play will win most of the time.

Let us also note that our argument did not depend at all on the number of piles: the first player can win in the manner described by the strategy described however many piles there are, provided the sums

$$a_m + b_m + c_m + \cdots + h_m, a_{m-1} + b_{m-1} + c_{m-1} + \cdots + h_{m-1}, \ldots,$$

$$a_1 + b_1 + c_1 + \cdots + h_1, a_0 + b_0 + c_0 + \cdots + h_0$$

are not all even.

Instead of counting to base 2 we could have used base 4, 8, etc. We give (without proof) the results when a system to base 4 is used. Suppose a, b, and c are of the forms

$$a = a_m 4^m + a_{m-1} 4^{m-1} + \cdots + a_1 4 + a_0,$$

$$b = b_m 4^m + b_{m-1} 4^{m-1} + \cdots + b_1 4 + b_0,$$

$$c = c_m 4^m + c_{m-1} 4^{m-1} + \cdots + c_1 4 + c_0,$$

respectively (where all the a_i's, etc., lie between 0 and 3). Then the initial position is a losing position if and only if every one of the sets $\{a_0, b_0, c_0\}$, $\{a_1 b_1 c_1\}, \ldots, \{a_m, b_m, c_m\}$ is equal to one of the sets $\{0,0,0\}$, $\{0,1,1\}$, $\{0,2,2\}$, $\{0,3,3\}$, $\{1,2,3\}$. If this is not the case, then the first player can win by correct play. He need only choose the matches he takes in such a way that afterwards all the sets *are* of this form. However, the easiest way of showing that this is always possible is to translate back to base 2.

131. The position of the game at any moment can be described by the ordered pair (x,y), where x is the number of matches in the first pile, and y is the number in the second pile. Thus the various possible positions can be represented geometrically by the points in the plane whose coordinates are nonnegative integers (fig. 74). When a player is at (x,y), he can move to any point on the horizontal line to the left of (x,y) (by taking matches from the first pile), or to any point on the vertical line below (x,y) (by taking matches from the second pile), or to any point of the diagonal line to the southwest of (x,y) (by taking the same number of matches from both piles).

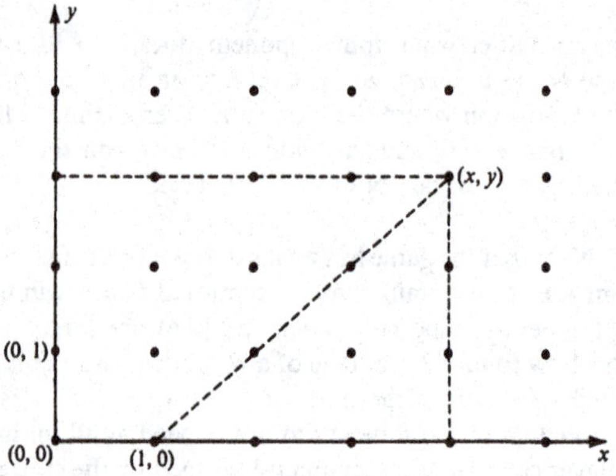

Fig. 74

We now proceed to analyze the game, using the method explained on page 17. In the present case properties (1), (2), and (3) can be stated as follows:

(1) (0,0) is a losing position.
(2) If P is a losing position, then every point on the horizontal to the right of P, the vertical above P, or the diagonal to the north-east of P, is winning.
(3) If every point on the horizontal to the left of P, the vertical below P, and the diagonal southeast of P is winning, then P is losing.

Applying property (2) with $P = (0,0)$, we see that the points (other than (0,0)) on the three lines drawn in fig. 75 are winning positions. By property (3), the positions (1,2) and (2,1) are losing. Another application of (2) shows that all uncircled points on the lines drawn in fig. 76 are winning positions.

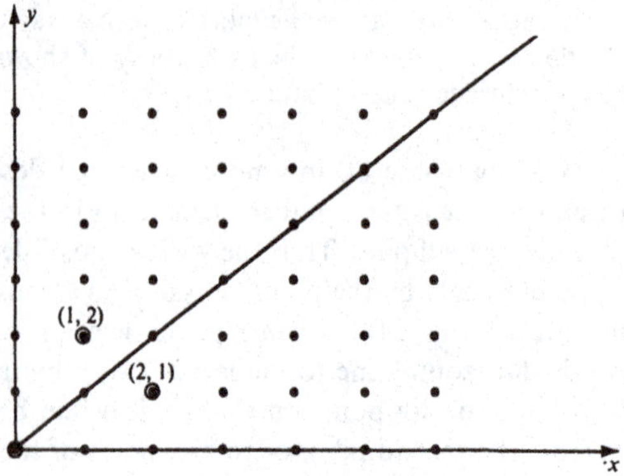

Fig. 75

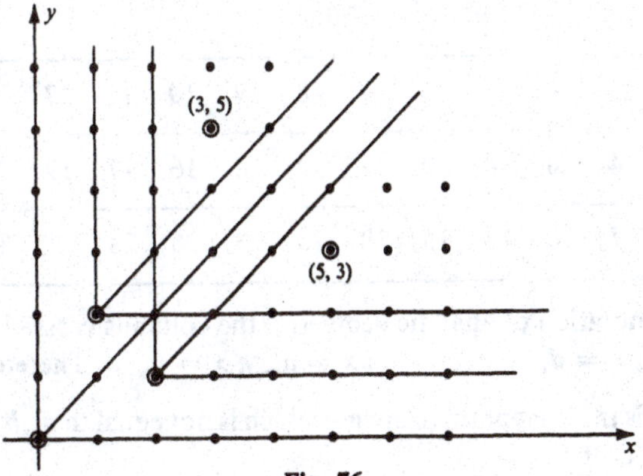

Fig. 76

Using property (3) again, we see that (3,5) and (5,3) are losing positions. Hence by (2), all uncircled points on the lines drawn in fig. 77 are winning positions. We now use (3) again to conclude that (4,7) and (7,4) are losing positions.

Continuing in this way, we obtain at the nth stage of the process two losing positions (a_n,b_n) and (b_n,a_n). The rule for finding (a_n,b_n) is the following:

Cross out the rows, columns, and 45° diagonals on which the points $(0,0)$, (a_1,b_1), (b_1,a_1), . . . , (a_{n-1},b_{n-1}), (b_{n-1},a_{n-1}) lie. Then (a_n,b_n) is the leftmost point of the diagonal $y = x + n$ which has not been crossed out.

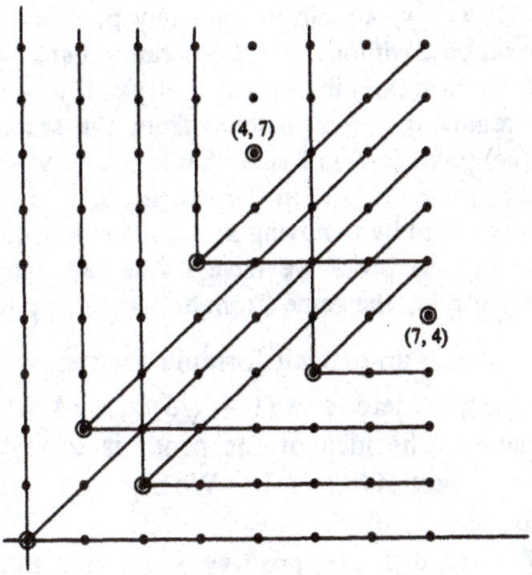

Fig. 77

Table 1

n	1	2	3	4	5	6	7	8	9	10	11	12	13	14	15
a_n	1	3	4	6	8	9	11	12	14	16	17	19	21	22	24
b_n	2	5	7	10	13	15	18	20	23	26	28	31	34	36	39

In the notation of analytic geometry, the columns crossed out are the lines $x = 0$, $x = a_1$, $x = b_1, \ldots, x = a_{n-1}$, $x = b_{n-1}$. Therefore

(i) a_n is the least positive integer which is not equal to $a_1, b_1, a_2, b_2, \ldots, a_{n-1}$, or b_{n-1}.
(ii) $b_n = a_n + n$.

Properties (i) and (ii) completely determine the sequence $\{(a_n,b_n)\}$, the first few terms of which are shown in table 1.

Remark. Our treatment has been designed to show how properties (i) and (ii) can be systematically derived from the general theory, as explained on page 17. If, however, one has somehow *guessed* (i) and (ii), it is easily verified that the pairs (a_n,b_n) and (b_n,a_n), together with (0,0), are indeed the losing positions. It suffices to show that the set $\mathfrak{L} = \{(0,0), (a_1,b_1), (b_1,a_1), (a_2,b_2), (b_2,a_2), \ldots\}$ satisfies the conditions of the theorem on page 19. Note first that every non-negative integer is the x coordinate of exactly one point in $\mathfrak{L}$ and is also the y coordinate of exactly one point in $\mathfrak{L}$. Moreover, *every* integer is the "coordinate difference" $y - x$ of exactly one pair (x,y) in $\mathfrak{L}$. In particular, no two distinct points P, P' of $\mathfrak{L}$ have the same x coordinate, y coordinate, or coordinate difference. This proves property (4) on page 19. To prove (5), suppose that (x,y) is not in $\mathfrak{L}$. If $x = y$, we can immediately play into the position (0,0). Therefore we may suppose without loss of generality that $x < y$. We know that there is a point of the form (x,z) in $\mathfrak{L}$ (where $z \neq y$). If $y > z$, we can play from (x,y) to (x,z) by removing $y - z$ matches from the second pile. If $y < z$, consider the (unique) point (a,b) in $\mathfrak{L}$ such that $b - a = y - x$. Since $b - a < z - x$, the pair (a,b) precedes (x,z) in the list $\{(a_n,b_n)\}$. Hence $a < x$, so one can play from (x,y) to (a,b) by removing $x - a$ matches from both piles.

In the course of this proof we have incidentally obtained an optimal strategy for actually playing the game (from any winning position).

We will now obtain an explicit formula for the pairs (a_n,b_n), namely $a_n = [n\tau]$, $b_n = [n\tau^2]$, where $\tau = (1 + \sqrt{5})/2$, and where the brackets denote integer parts. The idea of the proof is to show that the pairs $([n\tau], [n\tau^2])$ satisfy properties (i) and (ii). We begin by proving the following remarkable result.

Theorem. If α and β are positive irrational numbers such that $1/\alpha + 1/\beta = 1$, then every positive integer occurs once and only once among the numbers $[n\alpha]$ and $[n\beta]$, $n = 1, 2, 3, \ldots$.

Proof. Let S be the set consisting of the numbers $n\alpha$ and $n\beta$, $n = 1, 2, 3, \ldots$. Since α and β are irrational, the elements of S are also irrational. We want to show that there is exactly one element of S between 1 and 2, exactly one element between 2 and 3, etc. (For then, taking integer parts, we get every positive integer exactly once; see fig. 78.) In other words, we want to prove that if N is any positive integer, there are exactly $N - 1$ elements of S which are $< N$. Now $n\alpha < N$ for $n = 1, 2, \ldots, [N/\alpha]$, and $n\beta < N$ for $n = 1, 2, \ldots, [N/\beta]$. Hence the number of elements of S less than N is $[N/\alpha] + [N/\beta]$. We have $N/\alpha - 1 < [N/\alpha] < N/\alpha$, and $N/\beta - 1 < [N/\beta] < N/\beta$, since N/α and N/β are irrational. Adding these inequalities, we get $N/\alpha + N/\beta - 2 < [N/\alpha] + [N/\beta] < N/\alpha + N/\beta$. Since $1/\alpha + 1/\beta = 1$, this says that $N - 2 < [N/\alpha] + [N/\beta] < N$. Since $[N/\alpha] + [N/\beta]$ is an integer between $N - 2$ and N, it must $= N - 1$. This completes the proof.

Fig. 78

Now put $\alpha = \tau = (1 + \sqrt{5})/2$, and $\beta = \tau^2 = (3 + \sqrt{5})/2$. Since $\tau^2 = \tau + 1$, we have $1/\alpha + 1/\beta = 1/\tau + 1/\tau^2 = (\tau + 1)/\tau^2 = 1$. Since τ is irrational, our theorem tells us that every positive integer occurs exactly once among the numbers $[n\tau]$ and $[n\tau^2]$, $n = 1, 2, 3, \ldots$. This implies that $[n\tau]$ is the least positive integer not equal to $[\tau]$, $[\tau^2]$, $[2\tau]$, $[2\tau^2], \ldots, [(n - 1)\tau], [(n - 1)\tau^2]$. Moreover,

$$[n\tau^2] = [n(\tau + 1)] = [n\tau] + n.$$

Thus properties (i) and (ii) are satisfied by the pairs $([n\tau], [n\tau^2])$. Since these properties characterize the sequence $\{(a_n,b_n)\}$, we have

$$a_n = [n\tau], \qquad b_n = [n\tau^2].$$

This formula can be used to express the optimal strategy we found for winning the game in a curious form. We will merely state the rules and leave their verification to the reader. If θ is any real number, put $\{\theta\} = \theta - [\theta]$. We call $\{\theta\}$ the *fractional part* of θ; it satisfies $0 \leqq \{\theta\} < 1$. Suppose the initial position of the game is (x,y), where $x < y$. Then

(1) If $\{x\tau\} \leqq 1/\tau^2$, move to the position $(x, [x\tau] - x)$.
(2) If $\{x\tau\} > 1/\tau^2$ and $y > [x\tau] + 1$, move to $(x, [x\tau] + 1)$.
(3) If $\{x\tau\} > 1/\tau^2$ and $y < [x\tau] + 1$, move to $([(y - x)\tau], [(y - x)\tau^2])$.
(4) If $\{x\tau\} > 1/\tau^2$ and $y = [x\tau] + 1$, resign. You are in a losing position!

We wish now to give still another construction for the losing pairs, one which has the advantage that we can tell whether a given position is losing or not without having to calculate all previous losing pairs. For this purpose we use the number system known as the *F-system* (see pp. 15–16 for a description of it). Using the abbreviated notation, we list the F-expansions of the first few pairs (a_n,b_n).

n	a_n	b_n
1	1	10
2	100	1000
3	101	1010
4	1001	10010
5	10000	100000
6	10001	100010
7	10100	101000
8	10101	101010

We note that so far, every a_n ends in an even number of zeros (possibly none), and b_n is obtained from a_n by adding a zero at the end. We will now prove that these properties hold for *all* the pairs (a_n,b_n). For the moment, call (x,y) a *distinguished pair* if x ends in an even number of zeros, and $y = x0$ (that is, the F-expansion of y is obtained from that of x by adding a zero at the end). We wish to show that the distinguished pairs are precisely the pairs (a_n,b_n).

Every positive integer n belongs to exactly one distinguished pair. For if n ends in an even number of zeros, it belongs only to the pair $(n,n0)$, and if n ends in an odd number of zeros, it belongs only to the pair (m,n), where $n = m0$.

We show next that every positive integer z occurs as the difference $y - x$ of the members of a distinguished pair. Suppose

$$z = q_k q_{k-1} \cdots q_0 = q_k u_k + q_{k-1} u_{k-1} + \cdots + q_0 u_0.$$

If z ends with an odd number of zeros, let

$$x = z0 = q_k u_{k+1} + q_{k-1} u_k + \cdots + q_0 u_1$$
$$y = z00 = q_k u_{k+2} + q_{k-1} u_{k+1} + \cdots + q_0 u_2.$$

Then (x,y) is a distinguished pair, and

$$\begin{aligned} y - x &= q_k(u_{k+2} - u_{k+1}) + q_{k-1}(u_{k+1} - u_k) + \cdots + q_1(u_3 - u_2) \\ &= q_k u_k + q_{k-1} u_{k-1} + \cdots + q_0 u_0 = z. \end{aligned}$$

If, on the other hand, z ends in an even number of zeros, say $2m$, then

$$q_0 = q_2 = \cdots = q_{2m-1} = 0, \qquad q_{2m} = 1.$$

Let

$$x = q_k q_{k-1} \cdots q_{2m+1} \overbrace{0101 \cdots 01}^{2m+2 \text{ digits}},$$

$$y = x0 = q_k q_{k-1} \cdots q_{2m+1} \overbrace{0101 \cdots 010}^{2m+3 \text{ digits}}.$$

Then (x,y) is a distinguished pair. Moreover,

$$x = q_k u_{k+1} + q_{k-1} u_k + \cdots + q_{2m+1} u_{2m+2} + u_{2m} + u_{2m-2} + \cdots + u_2 + u_0,$$

$$y = q_k u_{k+2} + q_{k-1} u_{k+1} + \cdots + q_{2m+1} u_{2m+3} + u_{2m+1} + u_{2m-1} + \cdots + u_3 + u_1.$$

Hence

$$\begin{aligned} y - x &= q_k(u_{k+2} - u_{k+1}) + q_{k-1}(u_{k+1} - u_k) + \cdots + q_{2m+1}(u_{2m+3} - u_{2m+2}) \\ &\quad + (u_{2m+1} - u_{2m}) + (u_{2m-1} - u_{2m-2}) + \cdots + (u_3 - u_2) + (u_1 - u_0) \\ &= q_k u_k + q_{k-1} u_{k-1} + \cdots + q_{2m+1} u_{2m+1} + u_{2m-1} \\ &\quad + u_{2m-3} + \cdots + u_3 + u_1 + u_0 \end{aligned}$$

(noting that $u_1 - u_0 = 1 = u_0$). Now

$$\begin{aligned} &u_{2m-1} + u_{2m-3} + \cdots + u_3 + u_1 + u_0 \\ &= (u_{2m} - u_{2m-2}) + (u_{2m-2} - u_{2m-4}) \\ &\qquad + \cdots + (u_4 - u_2) + (u_2 - u_0) + u_0 \\ &= u_{2m}. \end{aligned}$$

Therefore

$$y - x = q_k u_k + q_{k-1} u_{k-1} + \cdots + q_{2m+1} u_{2m+1} + u_{2m} = z.$$

This shows that every positive integer z occurs at least once as the difference of the members of an ordered pair (x,y). Now suppose (x,y) and (x',y') are two distinguished pairs with $x < x'$. We claim that $y - x < y' - x'$. To prove this, let the F-expansions of x and x' be

$$x = q_k u_k + q_{k-1} u_{k-1} + \cdots + q_1 u_1 + q_0 u_0,$$
$$x' = q_k' u_k + q_{k-1}' u_{k-1} + \cdots + q_1' u_1 + q_0' u_0,$$

where if necessary we adjoin enough zeros at the beginning of x so that both x and x' have the same number of digits. Then if l is the greatest integer such that $q_l \neq q_l'$, we must have $q_l = 0$ and $q_l' = 1$. Note that $l \neq 1$, for then x' would end in one zero, contrary to the hypothesis that it ends in an even number of zeros. Hence $l \geqq 2$. Now

$$y = x0 = q_k u_{k+1} + q_{k-1} u_k + \cdots + q_1 u_2 + q_0 u_1,$$

and therefore

$$\begin{aligned} y - x &= q_k(u_{k+1} - u_k) + q_{k-1}(u_k - u_{k-1}) + \cdots + q_1(u_2 - u_1) \\ &\quad + q_0(u_1 - u_0) \\ &= q_k u_{k-1} + q_{k-1} u_{k-2} + \cdots + q_1 u_0 + q_0. \end{aligned}$$

Similarly,

$$y' - x' = q_k'u_{k-1} + q_{k-1}'u_{k-2} + \cdots + q_1'u_0 + q_0'.$$

Hence

$$(y' - x') - (y - x) = u_{l-1} + (q_{l-1}' - q_{l-1})u_{l-2} + \cdots + (q_1' - q_1)u_0 + (q_0' - q_0).$$

If l is odd, then

$$\begin{aligned}(y' - x') - (y - x) &\geqq u_{l-1} - (q_{l-1}u_{l-2} + \cdots + q_1u_0) - q_0\\ &\geqq u_{l-1} - (u_{l-2} + u_{l-4} + \cdots + u_1)\\ &= u_{l-1} - (u_{l-1} - 1) = 1,\end{aligned}$$

using equation (6) on page 16 (with $i = l - 1$).

If l is even, then we must have $q_i' = 1$ for at least one $i < l$; otherwise x' would end in an odd number of zeros. Hence in this case

$$\begin{aligned}(y' - x') - (y - x) &> u_{l-1} - (q_{l-1}u_{l-2} + \cdots + q_0u_0) - q_0\\ &\geqq u_{l-1} - (u_{l-2} + u_{l-4} + \cdots + u_0) - 1\\ &= u_{l-1} - (u_{l-1} - 1) - 1 = 0,\end{aligned}$$

using equation (5) on page 16 (with $i = l - 1$).

Thus in all cases $y' - x' > y - x$.

We are now ready to prove that the distinguished pairs are precisely the losing pairs (a_n,b_n). Denote the distinguished pairs by (x_1,y_1), (x_2,y_2), (x_3,y_3), . . . , where $x_1 < x_2 < x_3 < \cdots$. By what we have just shown, $y_1 - x_1 < y_2 - x_2 < y_3 - x_3 < \cdots$. But we saw earlier that every positive integer appears among the differences $y_n - x_n$. Therefore $y_1 - x_1 = 1$, $y_2 - x_2 = 2$, $y_3 - x_3 = 3$, and in general $y_n - x_n = n$ for every n (since otherwise the differences would skip some integer). From the fact that every positive integer occurs exactly once among the x's and y's, it follows that x_n is the least positive integer not equal to x_1, y_1, x_2, y_2, . . . , x_{n-1} or y_{n-1}. Thus the pairs (x_n,y_n) satisfy properties (i) and (ii). Since these properties characterize the pairs (a_n,b_n), we have $x_n = a_n$, $y_n = b_n$ for all n.

As an example of the use of this criterion for the losing pairs, let us determine whether the position (64,105) is winning or losing. The Fibonacci numbers $\leqq$ 105 are 1, 2, 3, 5, 8, 13, 21, 34, 55, 89. Therefore

$$64 = 55 + 8 + 1 = 100010001$$

$$105 = 89 + 13 + 3 = 1000100100.$$

We see from this that (64,105) is not distinguished and therefore is a winning position. Our earlier methods would require the calculation of the first 40 pairs (a_n,b_n) to obtain this result.

IX. POLYNOMIALS WITH MINIMUM DEVIATION FROM ZERO (TCHEBYCHEV POLYNOMIALS)

132. To obtain formulas **a** and **b**, we expand the left-hand side of De Moivre's formula

$$(\cos\alpha + i\sin\alpha)^n = \cos n\alpha + i\sin n\alpha$$

by the binomial theorem and then equate the real and imaginary parts of the resulting equation.

To prove **c** we divide the two sides of equation **a** by the corresponding sides of equation **b**, and then divide numerator and denominator of the quotient on the right-hand side of the resulting equation by $\cos^n\alpha$. Of course, we must assume that $\cos\alpha \neq 0$, but the formula is meaningless anyway when $\cos\alpha = 0$.

133. From problem 132**b** we have

$$\cos n\alpha = \cos^n\alpha - \binom{n}{2}\cos^{n-2}\alpha\sin^2\alpha + \binom{n}{4}\cos^{n-4}\alpha\sin^4\alpha - \cdots.$$

Let us now write

$$\alpha = \cos^{-1}x.$$

Then

$$\cos\alpha = x, \qquad \sin\alpha = \sqrt{1-x^2},$$

and therefore

$$\begin{aligned} T_n(x) &= \cos(n\cos^{-1}x) \\ &= x^n - \binom{n}{2}x^{n-2}(1-x^2) + \binom{n}{4}x^{n-4}(1-x^2)^2 - \cdots, \end{aligned}$$

so that $T(x)$ is indeed a polynomial of degree n. On removing the parentheses in this expression for $T_n(x)$, we find that the coefficient of x^n is

$$1 + \binom{n}{2} + \binom{n}{4} + \cdots,$$

which is equal to 2^{n-1} (see problem 58**a** of Volume 1).

Let us now find the roots of the equation

$$T_n(x) = 0.$$

Since $\cos\varphi = 0$ if and only if $\varphi = \frac{1}{2}(2k-1)\pi$ (k an integer), we have

$$T_n(x) = \cos(n\cos^{-1}x) = 0$$

if and only if

$$n\cos^{-1}x = \frac{1}{2}(2k-1)\pi, \qquad x = \cos\frac{1}{2n}(2k-1)\pi.$$

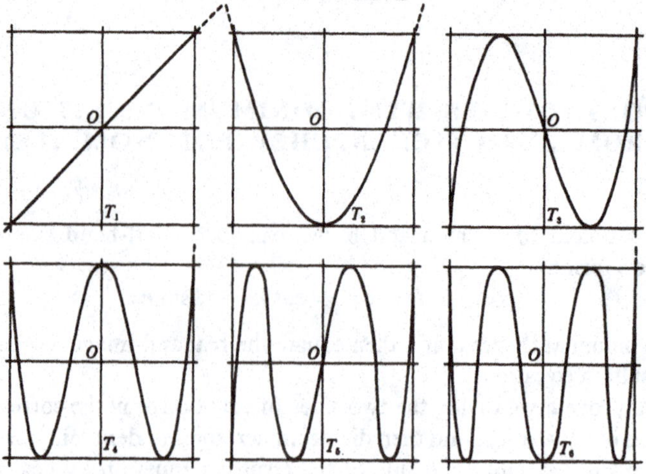

Fig. 79

By assigning the values 1, 2, 3, . . . , n to k we find that the n roots of the equation $T_n(x) = 0$ are

$$x_1 = \cos\frac{1}{2n}\pi, \quad x_2 = \cos\frac{3}{2n}\pi, \ldots, \quad x_n = \cos\frac{2n-1}{2n}\pi.$$

Next, it is clear that if x lies between -1 and $+1$, then $-1 \leqq \cos(n\cos^{-1}x) \leqq +1$ (since $-1 \leqq \cos\varphi \leqq 1$ for any φ). It is not difficult to find the values of x for which the polynomial $T_n(x) = \cos(n\cos^{-1}x)$ assumes the values $+1$ and -1. For $\cos\varphi = \pm 1$ if and only if $\varphi = k\pi$ (k an integer), and we deduce that

$$T_n(x) = \cos(n\cos^{-1}x) = \pm 1$$

if and only if

$$n\cos^{-1}x = k\pi, \qquad x = \frac{\cos k\pi}{n}.$$

Thus

$$T_n(\cos 0) = T_n(1) = 1, \qquad T_n\left(\cos\frac{\pi}{n}\right) = -1,$$
$$T_n\left(\cos\frac{2\pi}{n}\right) = 1, \qquad T_n\left(\cos\frac{3\pi}{n}\right) = -1, \ldots. \tag{1}$$

It is important to note that on the interval $-1 \leqq x \leqq 1$ the maximum and minimum values of $T_n(x)$ alternate. Thus at $x = 1$ the value of the polynomial is $+1$; it then drops to -1 (at the point $x = \cos\pi/n$), then again increases to $+1$ at the point $x = \cos 2\pi/n$, then drops again to -1 at the point $x = \cos 3\pi/n$, and so on. (See fig. 79, in which we show the first six Tchebychev polynomials.) The solution to problem 135 below is based on this property of $T_n(x)$.

134. It is easy to see that the polynomial

$$P(x) = x^2 + px + q = \left(x + \frac{p}{2}\right)^2 + q - \frac{p^2}{4}$$

assumes its minimum value $q - (p^2/4)$ when $x = -p/2$. We consider now two separate cases.

(1) $|-p/2| \geqq 1$ (fig. 80a). Suppose at $x = 1$ our polynomial assumes the value $P(1) = a$ and at $x = -1$ the value $P(-1) = b$; the deviation from zero of the polynomial P is then equal to the greater of $|a|$ and $|b|$. Consider now the polynomial

$$P_1(x) = x^2 + px + q - \frac{a+b}{2} = x^2 + px + q_1;$$

for this polynomial

$$P_1(1) = a - \frac{a+b}{2} = \frac{a-b}{2}, \qquad P_1(-1) = b - \frac{a+b}{2} = \frac{b-a}{2},$$

so that $P(1) = -P(-1)$. The deviation from zero of P_1 is not greater than the deviation of the original polynomial P. For the graph of P_1 is obtained from that of P by shifting it up or down such a distance that the two endpoints (where $x = 1$ and $x = -1$) are equally far from the x axis; see fig. 80a, where the graph of P_1 is shown dotted. Thus the deviation of P from zero is not less than that of P_1, and the deviation of P_1 is given by

$$\frac{|P_1(1) - P_1(-1)|}{2} = \frac{|(1 + p + q_1) - (1 - p + q_1)|}{2} = |p|.$$

Since by hypothesis $|p/2| \geqq 1$, it follows that the deviation of P is at least 2.

(2) $|-p/2| < 1$ (fig. 80b). Let us suppose that the point $x = -p/2$

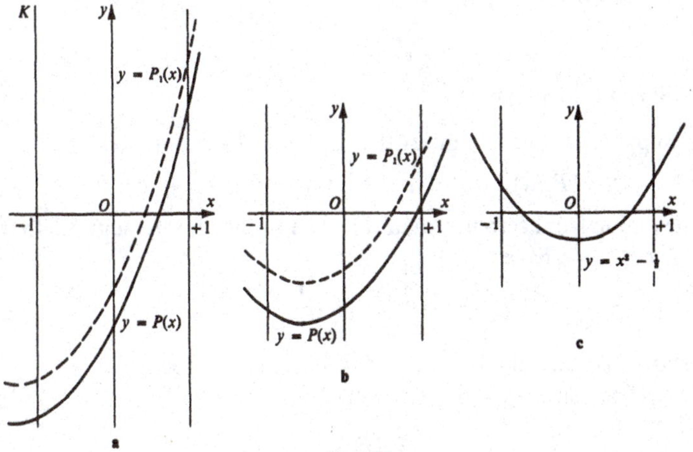

Fig. 80

is in the interval $-1 < -p/2 \leqq 0$, that is, $0 \leqq p/2 < 1$. (In the alternative case, $0 \leqq -p/2 < 1$, the argument needs only trivial alteration.) It is then clear that

$$\left|P(1) - P\left(-\frac{p}{2}\right)\right| \geq \left|P(-1) - P\left(-\frac{p}{2}\right)\right|$$

(see fig. 80b). As in case (1) we replace $P(x)$ by a new polynomial

$$P_1(x) = x^2 + px + q_1,$$

for which $P_1(-p/2) = -P_1(1)$. The graph of P_1 is obtained from that of P by shifting it vertically until the lowest point of the graph and the point with $x = 1$ are located at equal distances from the x axis (see fig. 90b). It is clear that the deviation from zero of P_1 is less than or equal to that of P. Hence the deviation of P is at least

$$\frac{|P_1(1) - P_1(-p/2)|}{2} = \frac{|(1 + p + q_1) - (p^2/4 - p^2/2 + q_1)|}{2}$$

$$= \frac{|p^2/4 + p + 1|}{2} = \frac{|(p/2 + 1)^2|}{2}.$$

Since $0 \leqq p/2 < 1$, this deviation is a minimum when $p = 0$, and for this value of p we deduce from

$$P_1\left(-\frac{p}{2}\right) = -P_1(1)$$

that

$$q_1 = -\frac{1}{2}.$$

We conclude that the quadratic polynomial with minimum deviation on the interval $-1 \leqq x \leqq +1$ is

$$P_0(x) = x^2 - \frac{1}{2}$$

(fig. 78c); its deviation is $\frac{1}{2}$.

135. Suppose that on the interval $-1 \leqq x \leqq 1$ the polynomial

$$P_n(x) = x^n + a_{n-1}x^{n-1} + \cdots + a_1x + a_0$$

has deviation from zero less than $(\frac{1}{2})^{n-1}$. In other words, suppose that for $-1 \leqq x \leqq 1$ we have

$$|P_n(x)| < \frac{1}{2^{n-1}}.$$

We show that this supposition leads to a contradiction.

Consider the polynomial R, where

$$R(x) = P_n(x) - \frac{1}{2^{n-1}} T_n(x).$$

Since the coefficients of the leading terms of P_n and $(\frac{1}{2})^{n-1}T_n$ are equal to 1, they cancel out on subtraction. The degree of R is therefore at most $n - 1$.

From (1) in the solution of problem 133, it follows that

$$R(1) = P_n(1) - \frac{1}{2^{n-1}} T_n(1) < 0,$$

$$R\left(\cos\frac{\pi}{n}\right) = P_n\left(\cos\frac{\pi}{n}\right) - \frac{1}{2^{n-1}} T_n\left(\cos\frac{\pi}{n}\right) > 0,$$

$$R\left(\cos\frac{2\pi}{n}\right) = P_n\left(\cos\frac{2\pi}{n}\right) - \frac{1}{2^{n-1}} T_n\left(\cos\frac{2\pi}{n}\right) < 0,$$

$$R\left(\cos\frac{3\pi}{n}\right) = P_n\left(\cos\frac{3\pi}{n}\right) - \frac{1}{2^{n-1}} T_n\left(\cos\frac{3\pi}{n}\right) > 0.$$

Thus $R(x)$ is negative at $x = 1$, whereas it is positive for $x = \cos \pi/n$. It follows (by the continuity of R), that there is a value x between 1 and $\cos \pi/n$ for which $R(x) = 0$. We see in exactly the same way that the equation $R(x) = 0$ has a root in each of the intervals $(\cos 2\pi/n, \cos \pi/n)$, $(\cos 3\pi/n, \cos 2\pi/n)$, $(\cos 4\pi/n, \cos 3\pi/n), \ldots, (-1, \cos (n-1)\pi/n)$.

We conclude that the equation $R(x) = 0$ has at least n roots. But R is of degree at most $n - 1$, and this is a contradiction unless R is the zero polynomial.[9] Hence $P_n = (\frac{1}{2})^{n-1}T_n$, contradicting the assumption that the deviation of P is less than $(\frac{1}{2})^{n-1}$ (for the deviation of T is *equal* to $(\frac{1}{2})^{n-1}$, as we saw in the solution to problem 133.) We have thus reached a contradiction, and the first part of the problem is solved.

We can prove the second part—that $(\frac{1}{2})^{n-1}T_n$ is the *only* polynomial of degree n with deviation equal to $(\frac{1}{2})^{n-1}$ on the interval $[-1,+1]$ by a refinement of the previous argument. Suppose P_n is a second such polynomial. As before, we consider the difference R, where

$$R(x) = P_n(x) - \frac{1}{2^{n-1}} T_n(x).$$

Here R is a polynomial of degree at most $n - 1$. At the points $x = 1$, $\cos 2\pi/n$, $\cos 4\pi/n, \ldots$ the values of R are $\leqq 0$, while at $x = \cos \pi/n$, $\cos 3\pi/n$, $\cos 5\pi/n$, $\ldots$ they are $\geqq 0$. It follows that in each of the intervals $[\cos \pi/n,\ 1]$, $[\cos 2\pi/n,\ \cos \pi/n]$, $[\cos 3\pi/n,\ \cos 2\pi/n]$, $\ldots$, $[-1, \cos (n-1)\pi/n]$ (possibly at their endpoints) $R(x)$ has at least one zero. We will show that in this case R has at least n zeros. It then follows as before that R is the zero polynomial, and therefore that $P_n = (\frac{1}{2})^{n-1}T_n$.

[9] If a polynomial $R(x)$ of degree $<n$ has n roots $x_1, x_2, x_3, \ldots, x_n$, then $R(x)$ is divisible by the polynomial $(x - x_1)(x - x_2) \cdots (x - x_n)$ of degree n. This is clearly impossible unless $R(x)$ is identically zero.

Consider the graph $y = R(x)$ near a point a at which $R(a) = 0$. This graph must either cross the x axis at the point $x = a$ or touch it. If the graph touches the axis, then $x = a$ is a double root of the equation $R(x) = 0$, in the sense that $(x - a)^2$ is a factor of $R(x)$. For the purposes of our argument such zeros must be counted twice.

Since a is a root of $R(x) = 0$, we can write $R(x) = (x - a)R_1(x)$ (by the factor theorem for polynomials).

If R does not have a double zero at $x = a$, then $R_1(a) \neq 0$. Since R_1 is continuous, this means that in a sufficiently small neighborhood of a, R_1 does not change sign. But as we pass from points on the left of $x = a$ to points on its right, the sign of the function $(x - a)$ *does* change, and so the sign of $(x - a)R_1(x)$ also changes. This means that the graph of $y = R(x)$ crosses the x axis at the point $x = a$, contrary to our hypothesis that it touches.

In proving that $R(x) = 0$ has n roots, we must count double roots twice. To show that R does have at least n roots, we prove the following more general assertion:

Let F be any continuous function defined on an interval $[a,b]$. Suppose this interval is divided into n smaller intervals

$$[a_n,a_{n-1}], [a_{n-1},a_{n-2}], \ldots, [a_2,a_1], [a_1,a_0] \qquad (a_n = a,\ a_0 = b).$$

Suppose that

$$F(a_i) \leqq 0 \qquad (i \text{ even}),$$

$$F(a_i) \geqq 0 \qquad (i \text{ odd}),$$

and that F has a zero in every interval (possibly at an end point). Then F has at least n zeros, provided double zeros are counted twice.

The proof is by mathematical induction. For $n = 1$ the theorem is trivial. Suppose we have proved it for all continuous functions and all intervals divided into fewer than n pieces. We then prove the result when there are n subintervals.

Consider first the case where n is even, so that $F(a_n) \leqq 0$. Suppose $F(a_n) < 0$. If $F(a_{n-1}) > 0$, then the graph $y = F(x)$ must cross the x axis at least once between $x = a_n$ and $x = a_{n-1}$. Thus F has at least one zero in the interval $[a_n,a_{n-1}]$ and, by the induction hypothesis, at least $n - 1$ zeros in the interval $[a_{n-1},a_0]$, a total of at least n zeros, as required.

Suppose next that $F(a_{n-1}) = 0$, but $F(a_{n-2}) \neq 0$. Then $F(a_{n-2}) < 0$. We know that F has at least $n - 2$ zeros on the interval $[a_{n-2},a_0]$, by the induction hypothesis. We must show that it has at least two zeros on the interval $[a_{n-1},a_{n-2}]$. We have one zero at a_{n-1}. If there are no other zeros to the left of a_{n-1}, then the graph of the curve remains under the x axis to the left of a_{n-1} (see fig. 79a and b). If the zero at a_{n-1} is not a double zero, then the graph is above the axis immediately to the right of a_{n-1}, and since

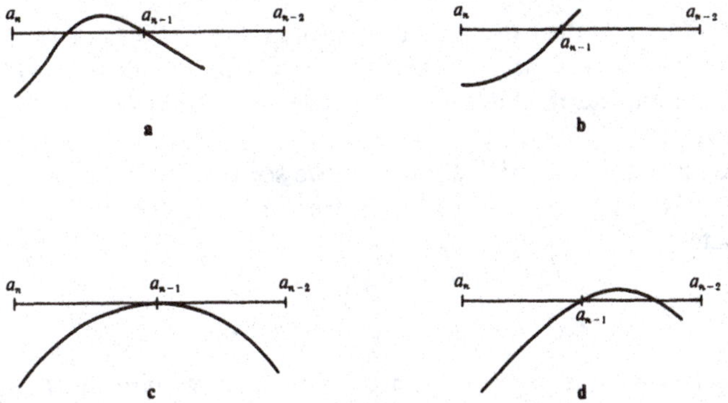

Fig. 81

it is below at a_{n-2}, it must cross the axis again somewhere in the interval $[a_{n-1},a_{n-2}]$ (see figs. 81c and d). Thus in every case there are at least two zeros (possibly a double zero at a_{n-1}) in the interval, as required.

Suppose more generally that

$$F(a_{n-1}) = F(a_{n-2}) = \cdots = F(a_{k+1}) = 0$$

but that

$$F(a_k) \neq 0.$$

By the induction hypothesis F has at least k zeros in the interval $[a_k,a_0]$. We must show that it has at least $n-k$ zeros in the interval $[a_n,a_k]$. If k is even, then $F(a_k) < 0$. We already have $n-k-1$ zeros $a_{n-1}, a_{n-2}, \ldots, a_{k+1}$. If there are no further zeros on the interval, then certainly all these zeros are simple (that is, not double) and there are no more zeros in the smaller interval $[a_n,a_{k+1}]$. But then the graph crosses the x axis $n-k-1$ times by the time it gets just to the right of a_{k+1}. Since $n-k-1$ is odd, the graph is above the x axis just to the right of a_{k+1}, and since it is below the x axis at a_k, it must cross the axis at least once in between. We thus have an $(n-k)$th zero (see fig. 82a).

The proof where k is odd is similar. The graph crosses the axis an even number of times if there are only $n-k-1$ simple zeros in the interval $[a_n,a_{k+1}]$, and is therefore below the x axis just to the right of a_{k+1}, whereas $F(a_k) > 0$. Hence there is a further zero in the last interval (see fig. 82b).

If there is no k for which $F(a_k) \neq 0$ $(k = n-1, n-2, \ldots, 1, 0)$, then we already have n zeros $a_{n-1}, a_{n-2}, \ldots, a_1, a_0$.

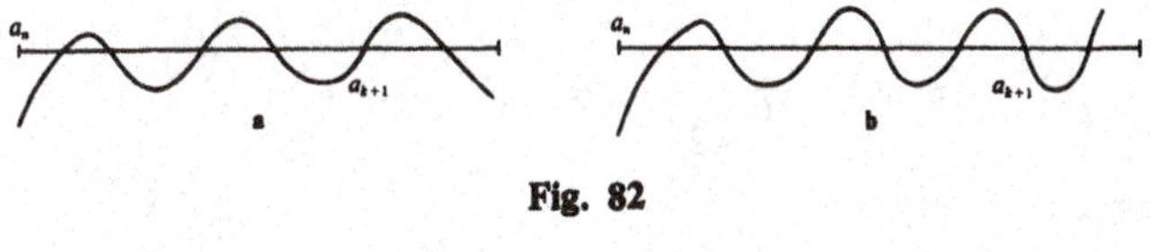

Fig. 82

If $F(a) = 0$, then by the induction hypothesis F has at least $n - 1$ zeros on the interval $[a_{n-1},a_0]$ and a_n is the required nth zero. Thus the proof is complete in this case. The case where n is odd is dealt with similarly.

Returning to the original problem we see that R satisfies all the conditions of the theorem and so has at least n zeros. The solution is now complete.

136. Let

$$P(x) = x^n + a_{n-1}x^{n-1} + \cdots + a_1 + a_0$$

be a polynomial of degree n whose deviation from zero on the interval $[-2,2]$ is equal to δ. Consider the polynomial

$$P_1(x) = P(2x) = (2x)^n + a_{n-1}(2x)^{n-1} + \cdots + a_1(2x) + a_0,$$

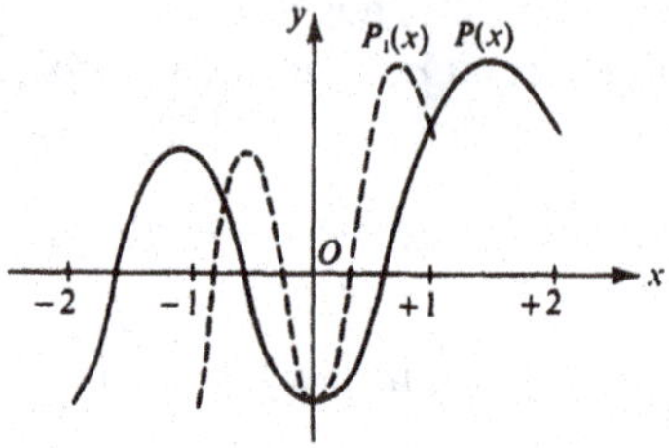

Fig. 83

whose graph is obtained from that of P by compressing it uniformly to half its width (fig. 83). The deviation of P_1 from zero on the interval $[-1,+1]$ is equal to the deviation δ of P on the interval $[-2,+2]$. Consider now the polynomial

$$\bar{P}(x) = \frac{1}{2^n} P_1(x)$$

$$= x^n + \frac{1}{2} a_{n-1}x^{n-1} + \frac{1}{4} a_{n-2}x^{n-2} + \cdots + \frac{1}{2^{n-1}} a_1 x + \frac{1}{2^n} a_0,$$

with leading coefficient 1. Its deviation on the interval $[-1,1]$ is clearly $\bar{\delta} = (\frac{1}{2})^n\delta$. But by the result of problem 135 we know that $\bar{\delta}$ is at least $(\frac{1}{2})^{n-1}$ and that $\bar{\delta} = (\frac{1}{2})^{n-1}$ only if $\bar{P} = (\frac{1}{2})^{n-1}T_n$. Therefore we see that $\delta = 2^n\bar{\delta}$ is at least 2 and is equal to 2 if and only if

$$P_1(x) = P(2x) = 2^n \frac{1}{2^{n-1}} T_n(x) = 2T_n(x)$$

$$P(x) = 2T_n\left(\frac{x}{2}\right) = 2 \cos \left(n \cos^{-1} \frac{x}{2}\right)$$

(where n is arbitrary).

Remark. It can be shown in exactly the same way that the deviation from zero of a polynomial P_n of degree n with leading coefficient 1 on the interval $[a,b]$ (where a and b are arbitrary real numbers with $a < b$) is at least $2\left(\frac{b-a}{4}\right)^n$, and is equal to $2\left(\frac{b-a}{4}\right)^n$ if and only if

$$P_n(x) = \left(\frac{b-a}{2}\right)^n \frac{1}{2^{n-1}} T_n\left[\frac{2}{b-a}(x-a) - 1\right]$$
$$= 2\left(\frac{b-a}{4}\right)^n T_n\left[\frac{2}{b-a}(x-a) - 1\right].$$

It follows, in particular, that in order for there to exist monic polynomials whose deviation on a given interval is less than any preassigned positive number (such as 1/1000 or 1/1,000,000) it is necessary and sufficient that the interval be of length less than 4.

137. This problem looks similar to the previous one but is solved quite differently. Denote the values assumed by a given polynomial

$$P(x) = x^n + a_{n-1}x^{n-1} + \cdots + a_1x + a_0$$

at the points $x = 0, 1, 2, \ldots, n$ by $P(0), P(1), P(2), \ldots, P(n)$, respectively. Consider now the polynomial Q, where

$$\begin{aligned} Q(x) = {} & P(0)\frac{(x-1)(x-2)(x-3)\cdots(x-n)}{(0-1)(0-2)(0-3)\cdots(0-n)} \\ & + P(1)\frac{(x-0)(x-2)(x-3)\cdots(x-n)}{(1-0)(1-2)(1-3)\cdots(1-n)} \\ & + P(2)\frac{(x-0)(x-1)(x-3)\cdots(x-n)}{(2-0)(2-1)(2-3)\cdots(2-n)} \\ & + \cdots \\ & + P(n)\frac{(x-0)(x-1)(x-2)\cdots[x-(n-1)]}{(n-0)(n-1)(n-2)\cdots[n-(n-1)]} \\ = {} & P(0)\frac{(x-1)(x-2)(x-3)\cdots(x-n)}{(-1)^n n!} \\ & + P(1)\frac{(x-0)(x-2)(x-3)\cdots(x-n)}{(-1)^{n-1}1!\,(n-1)!} \\ & + P(2)\frac{(x-0)(x-1)(x-3)\cdots(x-n)}{(-1)^{n-2}2!\,(n-2)!} \\ & + \cdots \\ & + P(n)\frac{(x-0)(x-1)(x-2)\cdots[x-(n-1)]}{n!}. \end{aligned}$$

The polynomials P and Q clearly have the same values at $x = 0, 1, 2, \ldots, n$. Thus $R(x) = P(x) - Q(x)$ has at least $n + 1$ zeros. Since $R(x)$ is a polynomial of degree at most n, it vanishes identically. Hence P and Q coincide. In particular, the leading coefficient of Q is 1, that is,

$$\frac{P(0)}{(-1)^n n!} + \frac{P(1)}{(-1)^{n-1}1!\,(n-1)!} + \frac{P(2)}{(-1)^{n-2}2!\,(n-2)!} + \frac{P(3)}{(-1)^{n-3}3!\,(n-3)!} + \cdots + \frac{P(n)}{n!} = 1. \quad (1)$$

Let us denote the deviation from zero of P on the set $\{0, 1, 2, \ldots, n\}$ by δ. Then the quantities $|P(0)|, |P(1)|, |P(2)|, \ldots, |P(n)|$ are $\leqq \delta$, and hence

$$\delta\left[\frac{1}{n!} + \frac{1}{1!\,(n-1)!} + \frac{1}{2!\,(n-2)!} + \frac{1}{3!\,(n-3)!} + \cdots + \frac{1}{n!}\right] \geqq 1.$$

Now

$$\frac{1}{n!} + \frac{1}{1!\,(n-1)!} + \frac{1}{2!\,(n-2)!} + \frac{1}{3!\,(n-3)!} + \cdots + \frac{1}{n!}$$

$$= \frac{1}{n!}\left[\binom{n}{0} + \binom{n}{1} + \binom{n}{2} + \binom{n}{3} + \cdots + \binom{n}{n-1} + \binom{n}{n}\right] = \frac{2^n}{n!}$$

(compare problem 57**a** of Volume 1). It follows that

$$\delta\frac{2^n}{n!} \geqq 1, \qquad \delta \geqq \frac{n!}{2^n}.$$

From this solution it also follows that there is a unique monic polynomial of degree n whose deviation from zero on the set $\{0, 1, 2, \ldots, n\}$ assumes the minimum value $n!/2^n$. It is the polynomial

$$P(x) = \frac{n!}{2^n}\left\{\frac{(x-1)(x-2)(x-3)\cdots(x-n)}{n!} + \frac{x(x-2)(x-3)\cdots(x-n)}{1!\,(n-1)!} + \frac{x(x-1)(x-3)\cdots(x-n)}{2!\,(n-2)!} + \cdots + \frac{x(x-1)(x-2)\cdots[x-(n-1)]}{n!}\right\},$$

for which

$$P(n) = -P(n-1) = P(n-2) = -P(n-3) = \cdots = (-1)^n P(0) = \frac{n!}{2^n}.$$

138. We use the geometric representation of complex numbers: the number

$$z = x + iy = r(\cos\theta + i\sin\theta)$$

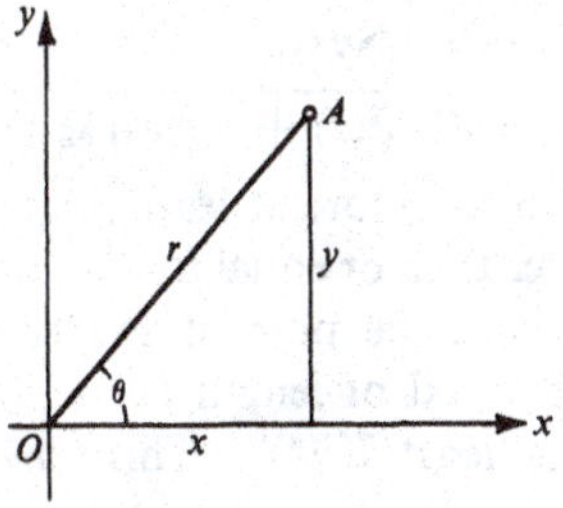

Fig. 84

is represented by the point of the plane with cartesian coordinates (x,y) and polar coordinates (r,θ) (fig. 84). If the points $A_1, A_2, \ldots, A_n$ of the plane correspond to the complex numbers $\alpha_1, \alpha_2, \ldots, \alpha_n$, and the point M to the variable complex number z, then

$$MA_1 = |z - \alpha_1|$$
$$MA_2 = |z - \alpha_2|$$
$$\cdots$$
$$MA_n = |z - \alpha_n|.$$

When complex numbers are multiplied, the absolute value of the product is equal to the product of the absolute values of the factors. Hence

$$MA_1 \cdot MA_2 \cdots MA_n = |(z - \alpha_1)(z - \alpha_2) \cdots (z - \alpha_n)|$$
$$= |z^n + a_{n-1}z^{n-1} + a_{n-2}z^{n-2} + \cdots + a_2z^2 + a_1z + a_0|,$$

where

$$z^n + a_{n-1}z^{n-1} + \cdots + a_1z + a_0 = P(z)$$

is the polynomial $(z - \alpha_1)(z - \alpha_2) \cdots (z - \alpha_n)$. In general, $P(z)$ has complex coefficients, and its roots are $\alpha_1, \alpha_2, \ldots, \alpha_n$.

It is now possible to see the connection between this problem and problem 136 on the deviation of polynomials from zero. Suppose we have selected our axes in the plane in such a way that the segment of length l is the section of the real axis between $-l/2$ and $+l/2$. Our problem can then be restated as follows: Given a polynomial P of degree n with leading coefficient 1 and complex coefficients, what is the minimum possible deviation of P on the real interval $I = [-l/2, +l/2]$? The deviation of a complex-valued function on an interval is defined as before; it is the maximum value of $|P(z)|$ on the interval. The only difference is that it is no longer possible to draw the graph of P.

We now write P in the form

$$P(z) = P_1(z) + iP_2(z),$$

where P_1 and P_2 have real coefficients. The coefficients of P_1 are the real parts of the corresponding coefficients of P, and those of P_2 are the imaginary parts. Hence P_1 is of degree n with leading coefficient 1, and

P_2 is of degree at most $n - 1$. Next,

$$|P(z)| = \sqrt{[P_1^2(z) + P_2^2(z)]} \geqq |P_1(z)|.$$

The absolute value of P is therefore at least as great as that of P_1, and so its deviation on I is greater than or equal to that of P_1. But by the result of problem 136 (see especially the note at the end of the solution), the deviation of P_1 on an interval of length l is at least $2(l/4)^n$. Hence the deviation of P on I is at least $2(l/4)^n$. This concludes the first part of the solution.

For the deviation of P on I to be equal to $2(l/4)^n$ it is necessary that $P_1(z) = 2(l/4)^n T_n(2z/l)$, and that P_2 vanishes at all the points where $|P_1|$ assumes its maximum [that is, $2(l/4)^n$]. But there are $n + 1$ such points

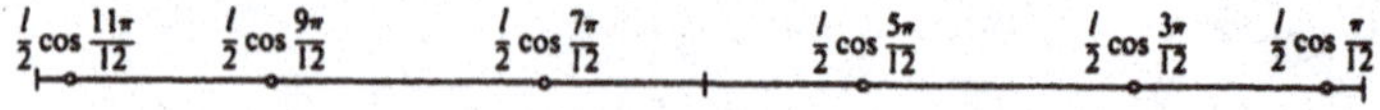

Fig. 85

(see the solutions to problems 133 and 135), and so the polynomial P_2 of degree $\leqq n - 1$ must vanish at $n + 1$ distinct points and must therefore be the zero polynomial (compare the solution of 135). We thus have

$$P(z) = 2\left(\frac{l}{4}\right)^n T_n\left(\frac{2z}{l}\right).$$

Interpreting this result in terms of the original formulation, we have shown that the product $MA_1 \cdot MA_2 \cdots MA_n$, where $A_1, A_2, \ldots, A_n$ are fixed points of the plane and M ranges over a segment I of length l, must assume a value of at least $2(l/4)^n$. And if it is to assume no larger value, the points $A_1, A_2, \ldots, A_n$ must be on I, and, moreover, in the positions of the roots of the equation $T_n(2z/l) = 0$ when we take I as part of the real axis and its midpoint as the origin. In other words, the distances of the points $A_1, A_2, \ldots, A_n$ from the center of I must be equal to

$$\frac{l}{2}\cos\frac{\pi}{2n},\ \frac{l}{2}\cos\frac{3\pi}{2n},\ \frac{l}{2}\cos\frac{5\pi}{2n},\ \ldots,\ \frac{l}{2}\cos\frac{(2n-1)\pi}{2n}$$

(fig. 85; see also the solution to problem 133).

X. FOUR FORMULAS FOR π

139a. Denoting the area of any plane figure F by $S(F)$, we have

$$S(\triangle OAM) = \tfrac{1}{2}OA \cdot MP = \tfrac{1}{2} \cdot 1 \cdot \sin\alpha = \tfrac{1}{2}\sin\alpha$$

$$S(\text{sector } OAM) = \frac{\alpha}{2\pi} S(\text{circle}) = \frac{\alpha}{2\pi} \cdot \pi$$

$$S(\triangle OAQ) = \tfrac{1}{2}OA \cdot AQ = \tfrac{1}{2} \cdot 1 \cdot \tan\alpha = \tfrac{1}{2}\tan\alpha.$$

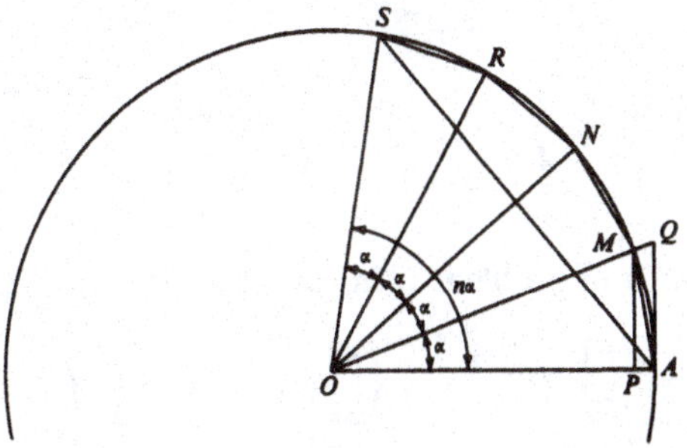

Fig. 86

The result now follows from the fact that

$$S(\triangle OAM) < S(\text{sector } OAM) < S(\triangle OAQ).$$

b. We have (fig. 86)

$$\alpha = 2S(\text{sector } OAM), \qquad \sin\alpha = 2S(\triangle OAM),$$
$$n\alpha = 2S(\text{sector } OAS), \qquad \sin n\alpha = 2S(\triangle OAS).$$

But clearly

$$\frac{\sin n\alpha}{n\alpha} = \frac{S(\triangle OAS)}{S(\text{sector } OAS)}$$
$$< \frac{S(\triangle OAM + \triangle OMN + \cdots + \triangle ORS)}{S(\text{sector } OAM + \text{sector } OMN + \cdots + \text{sector } ORS)}$$
$$= \frac{nS(\triangle OAM)}{nS(\text{sector } OAM)} = \frac{S(\triangle OAM)}{S(\text{sector } OAM)} = \frac{\sin\alpha}{\alpha}$$

as required.

140. Multiply the expression to be simplified by $\sin(\alpha/2^n)$. We obtain

$$\cos\frac{\alpha}{2}\cos\frac{\alpha}{4}\cdots\cos\frac{\alpha}{2^{n-1}}\left(\cos\frac{\alpha}{2^n}\sin\frac{\alpha}{2^n}\right)$$
$$= \frac{1}{2}\cos\frac{\alpha}{2}\cos\frac{\alpha}{4}\cdots\left(\cos\frac{\alpha}{2^{n-1}}\sin\frac{\alpha}{2^{n-1}}\right)$$
$$= \frac{1}{4}\cos\frac{\alpha}{2}\cos\frac{\alpha}{4}\cdots\left(\cos\frac{\alpha}{2^{n-2}}\sin\frac{\alpha}{2^{n-2}}\right)$$
$$\vdots$$
$$= \frac{1}{2^{n-2}}\cos\frac{\alpha}{2}\left(\cos\frac{\alpha}{4}\sin\frac{\alpha}{4}\right)$$
$$= \frac{1}{2^{n-1}}\left(\cos\frac{\alpha}{2}\sin\frac{\alpha}{2}\right) = \frac{1}{2^n}\sin\alpha,$$

whence

$$\cos\frac{\alpha}{2}\cos\frac{\alpha}{4}\cdots\cos\frac{\alpha}{2^n}=\frac{1}{2^n}\frac{\sin\alpha}{\sin\dfrac{\alpha}{2^n}}\qquad(\alpha\neq k\pi).$$

141a. The formula of problem 132**a** can be written in the form

$$\sin n\alpha=\sin^n\alpha\left[\binom{n}{1}\cot^{n-1}\alpha-\binom{n}{3}\cot^{n-3}\alpha+\binom{n}{5}\cot^{n-5}\alpha-\cdots\right].$$

Suppose now that $n = 2m + 1$ is odd. If α has any of the values

$$\frac{\pi}{2m+1},\ \frac{2\pi}{2m+1},\ \frac{3\pi}{2m+1},\ldots,\frac{m\pi}{2m+1},$$

then $\sin(2m+1)\alpha = 0$ and $\sin\alpha \neq 0$, so that

$$\binom{2m+1}{1}\cot^{2m}\alpha-\binom{2m+1}{3}\cot^{2m-2}\alpha+\binom{2m+1}{5}\cot^{2m-4}\alpha-\cdots=0.$$

We thus see that the equation

$$\binom{2m+1}{1}x^m-\binom{2m+1}{3}x^{m-1}+\binom{2m+1}{5}x^{m-2}\cdots=0$$

has the roots

$$\cot^2\frac{\pi}{2m+1},\ \cot^2\frac{2\pi}{2m+1},\ldots,\cot^2\frac{m\pi}{2m+1}.$$

b. We may rewrite the formula of 132c in the form

$$\frac{1}{\cot n\alpha}=\frac{\binom{n}{1}\dfrac{1}{\cot\alpha}-\binom{n}{3}\dfrac{1}{\cot^3\alpha}+\binom{n}{5}\dfrac{1}{\cot^5\alpha}-\cdots}{1-\binom{n}{2}\dfrac{1}{\cot^2\alpha}+\binom{n}{4}\dfrac{1}{\cot^4\alpha}-\binom{n}{6}\dfrac{1}{\cot^6\alpha}+\cdots}$$

or

$$\cot n\alpha=\frac{\cot^n\alpha-\binom{n}{2}\cot^{n-2}\alpha+\binom{n}{4}\cot^{n-4}\alpha-\cdots}{\binom{n}{1}\cot^{n-1}\alpha-\binom{n}{3}\cot^{n-3}\alpha+\binom{n}{5}\cot^{n-5}\alpha-\cdots}.$$

Suppose now that n is even. If α is equal to any of the numbers

$$\frac{\pi}{4n},\ \frac{5\pi}{4n},\ \frac{9\pi}{4n},\ldots,\frac{(4n-3)\pi}{4n},$$

then $\cot n\alpha = 1$; so for such values of α,

$$\frac{\cot^n \alpha - \binom{n}{2} \cot^{n-2} \alpha + \binom{n}{4} \cot^{n-4} \alpha - \cdots}{\binom{n}{1} \cot^{n-1} \alpha - \binom{n}{3} \cot^{n-3} \alpha + \binom{n}{5} \cot^{n-5} \alpha - \cdots} = 1.$$

It follows from this that the equation

$$x^n - \binom{n}{1}x^{n-1} - \binom{n}{2}x^{n-2} + \binom{n}{3}x^{n-3} + \binom{n}{4}x^{n-4} - \cdots = 0 \quad (1)$$

has the roots

$$\cot\frac{\pi}{4n},\quad \cot\frac{5\pi}{4n},\quad \cot\frac{9\pi}{4n},\ldots,\ \frac{\cot(4n-3)\pi}{4n}.$$

But

$$\frac{\cot(4n-3)\pi}{4n} = -\cot\frac{3\pi}{4n},$$

$$\cot\frac{(4n-7)\pi}{4n} = -\cot\frac{7\pi}{4n},\ldots,\ \cot\frac{(2n+1)\pi}{4n} = -\cot\frac{(2n-1)\pi}{4n},$$

and therefore the roots of equation (1) can also be written in the form

$$\cot\frac{\pi}{4n},\quad -\cot\frac{3\pi}{4n},\quad \cot\frac{5\pi}{4n},\quad -\cot\frac{7\pi}{4n},\ldots,\ \cot\frac{(2n-3)\pi}{4n},$$
$$-\cot\frac{(2n-1)\pi}{4n}.$$

c, d. The formulas of 132**a** and **b**, for $n = 2m$, can be written in the following form:

$$\sin 2m\alpha = \cos\alpha\sin\alpha\left[\binom{2m}{1}(1-\sin^2\alpha)^{m-1} - \binom{2m}{3}(1-\sin^2\alpha)^{m-2}\sin^2\alpha \right.$$
$$\left. + \binom{2m}{5}(1-\sin^2\alpha)^{m-3}\sin^4\alpha - \cdots\right]$$

and

$$\cos 2m\alpha = (1-\sin^2\alpha)^m - \binom{2m}{2}(1-\sin^2\alpha)^{m-1}\sin^2\alpha$$
$$+ \binom{2m}{4}(1-\sin^2\alpha)^{m-2}\sin^4\alpha - \cdots.$$

It follows in particular, as above, that the equations

$$\binom{2m}{1}(1-x)^{m-1} - \binom{2m}{3}(1-x)^{m-2}x + \binom{2m}{5}(1-x)^{m-3}x^2 - \cdots = 0$$

and

$$(1-x)^m - \binom{2m}{2}(1-x)^{m-1}x + \binom{2m}{4}(1-x)^{m-2}x^2 - \cdots = 0$$

have the roots

$$\sin^2\frac{2\pi}{m},\quad \sin^2\frac{2\pi}{2m},\quad \sin^2\frac{3\pi}{2m},\ldots,\sin^2\frac{(m-1)\pi}{2m},$$

and

$$\sin^2\frac{\pi}{4m},\quad \sin^2\frac{3\pi}{4m},\quad \sin^2\frac{5\pi}{4m},\ldots,\sin^2\frac{(2m-1)\pi}{4m},$$

respectively.

142a. Since the roots of the equation

$$\binom{2m+1}{1}x^m - \binom{2m+1}{3}x^{m-1} + \binom{2m+1}{5}x^{m-2} - \cdots = 0$$

are

$$\cot^2\frac{\pi}{2m+1},\quad \cot^2\frac{2\pi}{2m+1},\ldots,\cot^2\frac{m\pi}{2m+1}$$

(see problem 141**a**), the polynomial

$$\binom{2m+1}{1}x^m - \binom{2m+1}{3}x^{m-1} + \binom{2m+1}{5}x^{m-2} - \cdots$$

is divisible by each of

$$x-\cot^2\frac{\pi}{2m+1},\quad x-\cot^2\frac{2\pi}{2m+1},\ldots,x-\cot^2\frac{m\pi}{2m+1}.$$

We conclude that

$$\binom{2m+1}{1}x^m - \binom{2m+1}{3}x^{m-1} + \binom{2m+1}{5}x^{m-2} - \cdots$$

$$= A\left(x-\cot^2\frac{\pi}{2m+1}\right)\left(x-\cot^2\frac{2\pi}{2m+1}\right)\cdots\left(x-\cot^2\frac{m\pi}{2m+1}\right), \quad (1)$$

where A is a constant. On removing the parentheses on the right-hand side of this last equation and equating the coefficients of x^m and x^{m-1} on both sides, we find that

$$\binom{2m+1}{1} = A,$$

and

$$\binom{2m+1}{3} = A\left(\cot^2\frac{\pi}{2m+1} + \cot^2\frac{2\pi}{2m+1} + \cdots + \cot^2\frac{m\pi}{2m+1}\right).$$

Hence

$$\cot^2 \frac{\pi}{2m+1} + \cot^2 \frac{2\pi}{2m+1} + \cdots + \cot^2 \frac{m\pi}{2m+1}$$
$$= \frac{\binom{2m+1}{3}}{\binom{2m+1}{1}} = \frac{m(2m-1)}{3},$$

as required.

b. Since $\csc^2 \alpha = \cot^2 \alpha + 1$, we deduce from part **a** that

$$\csc^2 \frac{\pi}{2m+1} + \csc^2 \frac{2\pi}{2m+1} + \cdots + \csc^2 \frac{m\pi}{2m+1}$$
$$= \frac{m(2m-1)}{3} + m = \frac{2}{3} m(m+1).$$

c. It follows from the result of 141**b** that

$$x^n - \binom{n}{1} x^{n-1} - \binom{n}{2} x^{n-2} + \binom{n}{3} x^{n-3} + \cdots$$
$$= \left(x - \cot \frac{\pi}{4n}\right)\left(x + \cot \frac{3\pi}{4n}\right)\left(x - \cot \frac{5\pi}{4n}\right)\left(x + \cot \frac{7\pi}{4n}\right) \cdots$$
$$\left(x - \cot \frac{(2n-3)\pi}{4n}\right)\left(x + \cot \frac{(2n-1)\pi}{4n}\right).$$

Equating the coefficients of x^{n-1} on both sides of this last equation, we find that

$$\cot \frac{\pi}{4n} - \cot \frac{3\pi}{4n} + \cot \frac{5\pi}{4n} - \cot \frac{7\pi}{4n} + \cdots$$
$$+ \cot \frac{(2n-3)\pi}{4n} - \cot \frac{(2n-1)\pi}{4n} = \binom{n}{1} = n.$$

143. From the solution to 141**c** and **d** it follows that

$$\binom{2m}{1}(1-x)^{m-1} - \binom{2m}{3}(1-x)^{m-2}x + \binom{2m}{5}(1-x)^{m-3}x^2 - \cdots$$
$$= A\left(x - \sin^2 \frac{\pi}{2m}\right)\left(x - \sin^2 \frac{2\pi}{2m}\right) \cdots \left(x - \sin^2 \frac{(m-1)\pi}{2m}\right)$$

and

$$(1-x)^m - \binom{2m}{2}(1-x)^{m-1}x + \binom{2m}{4}(1-x)^{m-2}x^2 - \cdots$$
$$= B\left(x - \sin^2 \frac{\pi}{4m}\right)\left(x - \sin^2 \frac{3\pi}{4m}\right) \cdots \left(x - \sin^2 \frac{(2m-1)\pi}{4m}\right).$$

Now equating coefficients of the leading terms on both sides of these equations, we find that

$$A = (-1)^{m-1}\left[\binom{2m}{1} + \binom{2m}{3} + \binom{2m}{5} + \cdots\right],$$

$$B = (-1)^{m}\left[1 + \binom{2m}{2} + \binom{2m}{4} + \cdots\right],$$

whence (see problems 58**a** and **b** of Volume 1),

$$A = (-1)^{m-1}2^{2m-1}, \qquad B = (-1)^m 2^{2m-1}.$$

We next equate the constant terms in the same two equations. This yields

$$2m = \binom{2m}{1} = (-1)^{m-1}A \cdot \sin^2\frac{\pi}{2m}\sin^2\frac{2\pi}{2m}\cdots\sin^2\frac{(m-1)\pi}{2m},$$

$$1 = (-1)^m B\sin^2\frac{\pi}{4m}\sin^2\frac{3\pi}{4m}\sin^2\frac{5\pi}{4m}\cdots\sin^2\frac{(2m-1)\pi}{4m},$$

from which the required identities follow.

144a. Consider the identity

$$\cos\frac{\alpha}{2}\cos\frac{\alpha}{4}\cos\frac{\alpha}{8}\cdots\cos\frac{\alpha}{2^n} = \frac{1}{2^n}\frac{\sin\alpha}{\sin(\alpha/2^n)} \qquad (\alpha \neq 2^n k\pi)$$

obtained in problem 140. Take the limit as $n \to \infty$. We find

$$\cos\frac{\alpha}{2}\cos\frac{\alpha}{4}\cos\frac{\alpha}{8}\cdots = \lim_{n\to\infty}\left(\cos\frac{\alpha}{2}\cos\frac{\alpha}{4}\cos\frac{\alpha}{8}\cdots\cos\frac{\alpha}{2^n}\right)$$

$$= \sin\alpha\lim_{n\to\infty}\frac{1/2^n}{\sin(\alpha/2^n)} \qquad \left(\alpha \neq \lim_{n\to\infty}2^n k\pi\right).$$

But

$$\lim_{n\to\infty}\frac{\alpha/2^n}{\sin(\alpha/2^n)} = 1$$

because

$$\lim_{x\to 0}\frac{x}{\sin x} = 1;$$

this latter fact follows from the relation $1 \leq x/\sin x \leq 1/\cos x\,(0 < x < \pi/2)$, which in turn follows from the inequality of problem 139**a**. So

$$\sin\alpha\lim_{n\to\infty}\frac{1/2^n}{\sin(\alpha/2^n)} = \frac{\sin\alpha}{\alpha}\lim_{n\to\infty}\frac{\alpha/2^n}{\sin(\alpha/2^n)} = \frac{\sin\alpha}{\alpha} \qquad (\alpha \neq 0).$$

We thus obtain

$$\cos\frac{\alpha}{2}\cos\frac{\alpha}{4}\cos\frac{\alpha}{8}\cdots=\frac{\sin\alpha}{\alpha}\qquad(\alpha\neq 0)\tag{1}$$

and

$$\alpha=\frac{\sin\alpha}{\cos(\alpha/2)\cos(\alpha/4)\cos(\alpha/8)\cdots}\qquad(\alpha\neq k\pi).$$

To complete the argument we put $\alpha=\pi/2$ and use the relations $\sin(\pi/2)=1$, $\cos(\pi/2)=0$, and

$$\cos\frac{\theta}{2}=\sqrt{\left(\frac{1}{2}+\frac{1}{2}\cos\theta\right)}.$$

b. Substituting $\alpha=2\pi/3$, $\sin\alpha=\sqrt{3}/2$, $\cos(\alpha/2)=\frac{1}{2}$ in the identity (1), we obtain

$$\frac{1}{2}\sqrt{\left(\frac{1}{2}+\frac{1}{2}\cdot\frac{1}{2}\right)}\sqrt{\left[\frac{1}{2}+\frac{1}{2}\sqrt{\left(\frac{1}{2}+\frac{1}{2}\cdot\frac{1}{2}\right)}\right]}$$
$$\times\sqrt{\left\{\frac{1}{2}+\frac{1}{2}\sqrt{\left[\frac{1}{2}+\frac{1}{2}\sqrt{\left(\frac{1}{2}+\frac{1}{2}\cdot\frac{1}{2}\right)}\right]}\right\}}\cdots=\frac{3\sqrt{3}}{4\pi}.$$

145a. If the angle α lies in the first quadrant, then $\csc\alpha>\dfrac{1}{\alpha}>\cot\alpha$ (see problem 139**a**). It follows from the identities of problems 142**a** and **b** that

$$\frac{m(2m-1)}{3}<\left(\frac{2m+1}{\pi}\right)^2+\left(\frac{2m+1}{2\pi}\right)^2+\left(\frac{2m+1}{3\pi}\right)^2+\cdots$$
$$+\left(\frac{2m+1}{m\pi}\right)^2<\frac{m(2m+2)}{3},$$

or

$$\frac{\pi^2}{6}\left(1-\frac{1}{2m+1}\right)\left(1-\frac{2}{2m+1}\right)<1+\frac{1}{2^2}+\frac{1}{3^2}+\cdots+\frac{1}{m^2}$$
$$<\frac{\pi^2}{6}\left(1-\frac{1}{2m+1}\right)\left(1+\frac{1}{2m+1}\right).$$

Since the two outside terms of this double inequality both tend to the same limit $\pi^2/6$ as $m\to\infty$, we have

$$\lim_{m\to\infty}\left\{1+\frac{1}{2^2}+\frac{1}{3^2}+\cdots+\frac{1}{m^2}\right\}=\frac{\pi^2}{6}.$$

b. Let us find the sum of the squares of the roots of the equation in problem 141**a**. On equating the coefficient of x^{m-2} in the two sides of equation (1) in the solution to problem 142a, we find that the sum of all products of two distinct roots of this equation is

$$\frac{\binom{2m+1}{5}}{A} = \frac{\binom{2m+1}{5}}{\binom{2m+1}{1}} = \frac{m(2m-1)(m-1)(2m-3)}{30}.$$

Denoting the roots by $\alpha, \beta, \gamma, \ldots$, we have the identity

$$\alpha^2 + \beta^2 + \gamma^2 + \cdots = (\alpha + \beta + \gamma \cdots)^2 - 2(\alpha\beta + \alpha\gamma + \cdots + \beta\gamma + \cdots)$$

where the last sum is taken over all possible distinct pairs. Applying this formula, we find that

$$\cot^4 \frac{\pi}{2m+1} + \cot^4 \frac{2\pi}{2m+1} + \cot^4 \frac{3\pi}{2m+1} + \cdots + \cot^4 \frac{m\pi}{2m+1}$$
$$= \frac{m^2(2m-1)^2}{9} - \frac{m(2m-1)(m-1)(2m-3)}{15}$$
$$= \frac{m(2m-1)(4m^2+10m-9)}{45}.$$

Next,

$$\csc^4 \alpha = (\cot^2 \alpha + 1)^2 = \cot^4 \alpha + 2\cot^2 \alpha + 1,$$

so that

$$\csc^4 \frac{\pi}{2m+1} + \csc^4 \frac{2\pi}{2m+1} + \cdots + \csc^4 \frac{m\pi}{2m+1}$$
$$= \frac{m(2m-1)(4m^2+10m-9)}{45} + \frac{2m(2m-1)}{3} + m$$
$$= \frac{8m(m+1)(m^2+m+3)}{45}.$$

Now, as in the solution to part a, we obtain the double inequality

$$\frac{m(2m-1)(4m^2+10m-9)}{45} < \left(\frac{2m+1}{\pi}\right)^4 + \left(\frac{2m+1}{2\pi}\right)^4$$
$$+ \left(\frac{2m+1}{3\pi}\right)^4 + \cdots + \left(\frac{2m+1}{m\pi}\right)^4 < \frac{8m(m+1)(m^2+m+3)}{45}$$

or

$$\frac{\pi^2}{90}\left(1 - \frac{1}{2m+1}\right)\left(1 - \frac{2}{2m+1}\right)\left(1 + \frac{3}{2m+1} - \frac{13}{(2m+1)^2}\right)$$
$$< 1 + \frac{1}{2^4} + \frac{1}{3^4} + \cdots + \frac{1}{m^4} < \frac{\pi^4}{90}\left(1 - \frac{1}{(2m+1)^2}\right)\left(1 + \frac{11}{(2m+1)^2}\right),$$

from which we deduce by letting $m \to \infty$ that

$$1 + \frac{1}{2^4} + \frac{1}{3^4} + \frac{1}{4^4} + \cdots = \frac{\pi^4}{90}.$$

Remark. We may evaluate the sum of the cubes, fourth powers, and so on, of the roots of the equation in problem **141a** in the same manner, and we can deduce the following sequence of formulas, first given by Euler:

$$1 + \frac{1}{2^6} + \frac{1}{3^6} + \frac{1}{4^6} + \cdots = \frac{\pi^6}{945}$$

$$1 + \frac{1}{2^8} + \frac{1}{3^8} + \frac{1}{4^8} + \cdots = \frac{\pi^8}{9450}$$

$$1 + \frac{1}{2^{10}} + \frac{1}{3^{10}} + \frac{1}{4^{10}} + \cdots = \frac{\pi^{10}}{93{,}555}$$

$$1 + \frac{1}{2^{12}} + \frac{1}{3^{12}} + \frac{1}{4^{12}} + \cdots = \frac{691\pi^{12}}{638{,}512{,}875}.$$

146a. We have the identity

$$\cot \alpha - \cot \beta = \frac{\sin(\beta - \alpha)}{\sin \alpha \sin \beta} = \sin(\beta - \alpha) \csc \alpha \csc \beta.$$

Using the identity of problem **142c** we deduce from it that

$$\sin \frac{\pi}{2n}\left(\csc \frac{\pi}{4n} \csc \frac{3\pi}{4n} + \csc \frac{5\pi}{4n} \csc \frac{7\pi}{4n} + \cdots + \csc \frac{(2n-3)\pi}{4n} \csc \frac{(2n-1)\pi}{4n}\right) = n,$$

or

$$\csc \frac{\pi}{4n} \csc \frac{3\pi}{4n} + \csc \frac{5\pi}{4n} \csc \frac{7\pi}{4n} + \cdots + \csc \frac{(2n-3)\pi}{4n} \csc \frac{(2n-1)\pi}{4n} = \frac{n}{\sin(\pi/2n)}.$$

On the other hand,

$$\cot \alpha - \cot \beta = \tan(\beta - \alpha)(1 + \cot \alpha \cot \beta),$$

since $\cot(\beta - \alpha) = (1 + \cot \alpha \cot \beta)/(\cot \alpha - \cot \beta)$. It follows from the same identity that

$$\tan \frac{\pi}{2n}\left[\cot \frac{\pi}{4n} \cot \frac{3\pi}{4n} + \cot \frac{5\pi}{4n} \cot \frac{7\pi}{4n} + \cdots + \cot \frac{(2n-3)\pi}{4n} \cot \frac{(2n-1)\pi}{4n} + \frac{n}{2}\right] = n,$$

or

$$\cot\frac{\pi}{4n}\cot\frac{3\pi}{4n}+\cot\frac{5\pi}{4n}\cot\frac{7\pi}{4n}+\cdots+\cot\frac{(2n-3)\pi}{4n}\cot\frac{(2n-1)\pi}{4n}$$
$$=\frac{n}{\tan(\pi/2n)}-\frac{n}{2}.$$

Proceeding as in problem 145**a** we obtain the double inequality

$$\frac{n}{\sin(\pi/2n)}>\frac{4n}{\pi}\cdot\frac{4n}{3\pi}+\frac{4n}{5\pi}\cdot\frac{4n}{7\pi}+\cdots+\frac{4n}{(2n-3)\pi}\cdot\frac{4n}{(2n-1)\pi}$$
$$>\frac{n}{\tan(\pi/2n)}-\frac{n}{2},$$

or

$$\frac{\pi^2}{8n}\frac{1}{\sin(\pi/2n)}>\frac{2}{1\cdot 3}+\frac{2}{5\cdot 7}+\cdots+\frac{2}{(2n-3)(2n-1)}$$
$$>\frac{\pi^2}{8n}\frac{1}{\tan(\pi/2n)}-\frac{\pi^2}{16n},$$

or, finally,

$$\frac{\pi}{4}\frac{\pi/2n}{\sin(\pi/2n)}>1-\frac{1}{3}+\frac{1}{5}-\frac{1}{7}+\cdots+\frac{1}{2n-3}-\frac{1}{2n-1}$$
$$>\frac{\pi}{4}\left(\frac{\pi/2n}{\tan(\pi/2n)}-\frac{\pi}{4n}\right).$$

As $n\to\infty$ the two outside terms of this double inequality tend to the same limit $\pi/4$. This is because

$$\lim_{\alpha\to\infty}\frac{\alpha}{\sin\alpha}=1,\qquad \lim_{\alpha\to 0}\frac{\alpha}{\tan\alpha}=\lim_{\alpha\to 0}\left(\frac{\alpha}{\sin\alpha}\cdot\cos\alpha\right)=1.$$

(See the solution of 144**a**.) It follows that

$$1-\frac{1}{3}+\frac{1}{5}-\frac{1}{7}+\cdots=\frac{\pi}{4},$$

as required.

b. Let us find the sum of the squares of the roots of the equation in problem 141**b**. As in problem 145**b** we see that the sum of the products of the roots taken two at a time is $-\binom{n}{2}=-n(n-1)/2$, and that the sum of the squares of the roots is $n^2+n(n-1)=n(2n-1)$.

We deduce from the resulting identity

$$\cot^2\frac{\pi}{4n}+\cot^2\frac{3\pi}{4n}+\cot^2\frac{5\pi}{4n}+\cdots+\cot^2\frac{(2n-1)\pi}{4n}=n(2n-1),$$

that

$$\csc^2\frac{\pi}{4n}+\csc^2\frac{3\pi}{4n}+\csc^2\frac{5\pi}{4n}+\cdots+\csc^2\frac{(2n-1)\pi}{4n}$$
$$=n(2n-1)+n=2n^2,$$

from which, as in the solution to 145**a**, we conclude that

$$n(2n-1)<\left(\frac{4n}{\pi}\right)^2+\left(\frac{4n}{3\pi}\right)^2+\left(\frac{4n}{5\pi}\right)^2+\cdots+\left(\frac{4n}{(2n-1)\pi}\right)^2<2n^2,$$

or

$$\frac{\pi^2}{8}\left(1-\frac{1}{2n}\right)<1+\frac{1}{3^2}+\frac{1}{5^2}+\cdots+\frac{1}{(2n-1)^2}<\frac{\pi^2}{8},$$

and so that

$$1+\frac{1}{3^2}+\frac{1}{5^2}+\frac{1}{7^2}+\cdots=\frac{\pi^2}{8}.$$

This last formula is in fact equivalent to Euler's formula; for

$$1+\frac{1}{3^2}+\frac{1}{5^2}+\frac{1}{7^2}+\cdots$$
$$=\left(1+\frac{1}{2^2}+\frac{1}{3^2}+\frac{1}{4^2}+\cdots\right)-\left(\frac{1}{2^2}+\frac{1}{4^2}+\frac{1}{6^2}+\frac{1}{8^2}+\cdots\right)$$
$$=\left(1+\frac{1}{2^2}+\frac{1}{3^2}+\frac{1}{4^2}+\cdots\right)-\frac{1}{4}\left(1+\frac{1}{2^2}+\frac{1}{3^2}+\frac{1}{4^2}+\cdots\right)$$
$$=\frac{\pi^2}{6}-\frac{1}{4}\cdot\frac{\pi^2}{6}=\frac{\pi^2}{8}.$$

Remark. By determining the sums of the cubes, fourth powers, and so on, of the roots of the equation of problem 132**b** we may obtain the following sequence of formulas:

$$1-\frac{1}{3^3}+\frac{1}{5^3}-\frac{1}{7^3}+\cdots=\frac{\pi^3}{32},$$

$$1+\frac{1}{3^4}+\frac{1}{5^4}+\frac{1}{7^4}+\cdots=\frac{\pi^4}{96},$$

$$1-\frac{1}{3^5}+\frac{1}{5^5}-\frac{1}{7^5}+\cdots=\frac{5\pi^5}{1536},$$

$$1+\frac{1}{3^6}+\frac{1}{5^6}+\frac{1}{7^6}+\cdots=\frac{\pi^6}{960}.$$

These formulas were also first discovered by Euler.

147. We consider the following two expressions:

$$\frac{\sin(2\pi/4m)}{\sin(\pi/4m)}\,\frac{\sin(2\pi/4m)}{\sin(3\pi/4m)}\,\frac{\sin(4\pi/4m)\sin(4\pi/4m)}{\sin(3\pi/4m)\sin(5\pi/4m)}$$

$$\cdots\frac{\sin((2m-2)\pi/4m)\sin((2m-2)\pi4/m)}{\sin((2m-3)\pi/4m)\sin((2m-1)\pi/4m)} \qquad (1)$$

and

$$\frac{\sin(2\pi/4m)\sin(4\pi/4m)}{\sin(3\pi/4m)\sin(3\pi/4m)}\,\frac{\sin(4\pi/4m)\sin(6\pi/4m)}{\sin(5\pi/4m)\sin(5\pi/4m)}$$

$$\cdots\frac{\sin((2m-2)\pi/4m)\sin(2m\pi/4m)}{\sin((2m-1)\pi/4m)\sin((2m-1)\pi/4m)}. \qquad (2)$$

The consideration of these two expressions is suggested by the form of Wallis's formula. By the identity of problem 143 these two expressions are equal respectively to

$$\left(\frac{\sqrt{m}/2^{m-1}}{\sqrt{2}/2^m}\right)^2\sin\frac{\pi}{4m}\sin\frac{(2m-1)\pi}{4m}=2m\sin\frac{\pi}{4m}\cos\frac{\pi}{4m}=m\sin\frac{\pi}{2m}$$

and

$$\left(\frac{\sqrt{m}/2^{m-1}}{\sqrt{2}/2^m}\right)^2\frac{\sin(2m\pi/4m)}{\sin(2\pi/4m)}\sin^2\frac{\pi}{4m}$$

$$=2m\frac{\sin^2(\pi/4m)}{2\sin(\pi/4m)\cos(\pi/4m)}=m\tan\frac{\pi}{4m}.$$

We now show that the quantity (1) is decreased by substituting angles for their sines throughout, and, similarly, that (2) is increased by this substitution. We use the fact that

$$\begin{aligned}\sin^2 k\alpha-\sin^2\alpha&=(\sin k\alpha+\sin\alpha)(\sin k\alpha-\sin\alpha)\\&=2\sin\frac{(k+1)\alpha}{2}\cos\frac{(k-1)\alpha}{2}\cdot 2\sin\frac{(k-1)\alpha}{2}\cos\frac{(k+1)\alpha}{2}\\&=2\sin\frac{(k-1)\alpha}{2}\cos\frac{(k-1)\alpha}{2}\cdot 2\sin\frac{(k+1)\alpha}{2}\cos\frac{(k+1)\alpha}{2}\\&=\sin(k-1)\alpha\sin(k+1)\alpha.\end{aligned}$$

It follows from this that

$$\frac{\sin(k-1)\alpha}{\sin k\alpha}\,\frac{\sin(k+1)\alpha}{\sin k\alpha}=\frac{\sin^2 k\alpha-\sin^2\alpha}{\sin^2 k\alpha}=1-\left(\frac{\sin\alpha}{\sin k\alpha}\right)^2,$$

while

$$\frac{(k-1)\alpha}{k\alpha}\,\frac{(k+1)\alpha}{k\alpha}=\frac{(k^2-1)\alpha^2}{(k\alpha)^2}=1-\left(\frac{\alpha}{k\alpha}\right)^2.$$

From the result of problem 139**b**, $\sin\alpha/\sin k\alpha > \alpha/k\alpha$ and therefore

$$1-\left(\frac{\sin\alpha}{\sin k\alpha}\right)^2 < 1-\left(\frac{\alpha}{k\alpha}\right)^2.$$

As a consequence of this,

$$\frac{\sin(k-1)\alpha}{\sin k\alpha}\,\frac{\sin(k+1)\alpha}{\sin k\alpha} < \frac{(k-1)\alpha}{k\alpha}\,\frac{(k+1)\alpha}{k\alpha},$$

$$\frac{\sin k\alpha}{\sin(k-1)\alpha}\,\frac{\sin k\alpha}{\sin(k+1)\alpha} > \frac{k\alpha}{(k-1)\alpha}\,\frac{k\alpha}{(k+1)\alpha}.$$

We thus have

$$\frac{2\pi/4m}{\pi/4m}\cdot\frac{2\pi/4m}{3\pi/4m}\cdot\frac{4\pi/4m}{3\pi/4m}\cdot\frac{4\pi/4m}{5\pi/4m}$$
$$\cdots\frac{(2m-2)\pi/4m}{(2m-3)\pi/4m}\,\frac{(2m-2)\pi/4m}{(2m-1)\pi/4m} < m\sin\frac{\pi}{2m}$$

and

$$\frac{2\pi/4m}{3\pi/4m}\cdot\frac{4\pi/4m}{3\pi/4m}\cdot\frac{4\pi/4m}{5\pi/4m}\cdot\frac{6\pi/4m}{5\pi/4m}$$
$$\cdots\frac{(2m-2)\pi/4m}{(2m-1)\pi/4m}\cdot\frac{2m\pi/4m}{(2m-1)\pi/4m} > m\tan\frac{\pi}{4m}$$

or

$$\frac{2}{1}\cdot\frac{2}{3}\cdot\frac{4}{3}\cdot\frac{4}{5}\cdots\frac{2m-2}{2m-3}\cdot\frac{2m-2}{2m-1} < m\sin\frac{\pi}{2m}$$

and

$$\frac{2}{3}\cdot\frac{4}{3}\cdot\frac{4}{5}\cdot\frac{6}{5}\cdots\frac{2m-2}{2m-1}\cdot\frac{2m}{2m-1} > m\tan\frac{\pi}{4m}.$$

The last two inequalities can be combined in the following double inequality:

$$m\sin\frac{\pi}{2m} > \frac{2}{1}\cdot\frac{2}{3}\cdot\frac{4}{3}\cdot\frac{4}{5}\cdots\frac{2m-2}{2m-3}\cdot\frac{2m-2}{2m-1} > \frac{2m-1}{2m}\,m\tan\frac{\pi}{4m},$$

or

$$\frac{\pi}{2}\,\frac{\sin(\pi/2m)}{\pi/2m} > \frac{2}{1}\cdot\frac{2}{3}\cdot\frac{4}{3}\cdot\frac{4}{5}\cdots\frac{2m-2}{2m-3}\cdot\frac{2m-2}{2m-1} > \frac{\pi}{2}\left(1-\frac{1}{2m}\right)\frac{\tan(\pi/4m)}{\pi/4m}.$$

The two outside terms of this double inequality tend to the same limit $\pi/2$ as $m\to\infty$. (See the solution to problem 146**a**.) It follows that

$$\frac{2}{1}\cdot\frac{2}{3}\cdot\frac{4}{3}\cdot\frac{4}{5}\cdot\frac{6}{5}\cdot\frac{6}{7}\cdots = \frac{\pi}{2}.$$

XI. THE CALCULATION OF AREAS OF REGIONS BOUNDED BY CURVES

148. Divide the segment OD of length a into n equal parts and construct on each small segment $M_{k-1}M_k$ a rectangle of the type described on p. 26 (fig. 87). The base of each rectangle is a/n, while the height of the kth rectangle is $(ka/n)^2$. The total area of all the rectangles is therefore

$$S_n = \frac{a}{n}\left[\left(\frac{a}{n}\right)^2 + \left(2\frac{a}{n}\right)^2 + \cdots + \left(n\frac{a}{n}\right)^2\right]$$
$$= \frac{a^3}{n^3}(1^2 + 2^2 + \cdots + n^2).$$

It is well known (and can be proved by induction, as in the footnote on page 105 of Volume 1) that

$$1^2 + 2^2 + \cdots + n^2 = \frac{n(n+1)(2n+1)}{6},$$

and so

$$S_n = \frac{a^3}{n^3}\,\frac{n(n+1)(2n+1)}{6}.$$

The area S of the region bounded by the parabola is equal to the limit of S_n as the number of rectangles increases indefinitely. That is,

$$S = \lim_{n\to\infty} S_n = \lim_{n\to\infty} a^3\frac{(2n^3 + 3n^2 + n)}{6n^3}$$
$$= \lim_{n\to\infty} a^3\left(\frac{1}{3} + \frac{1}{2n} + \frac{1}{6n^2}\right) = \frac{a^3}{3}.$$

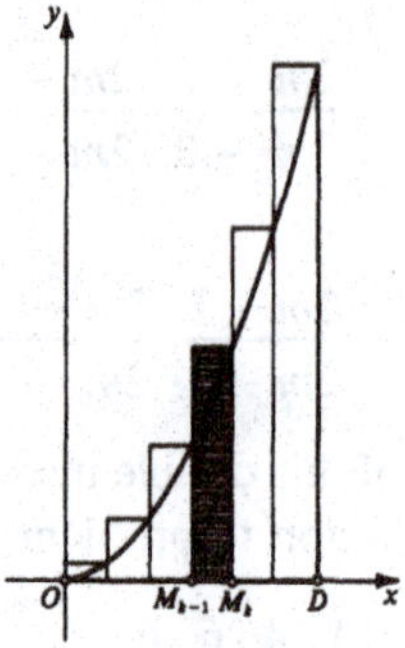

Fig. 87

149a. The required area is 2π. The proof follows at once from fig. 88a, from which we see that the area is that of $AOBDC$, and that this figure has the same area as the rectangle $PQRS$. The two portions of the area which do not lie in the rectangle are congruent to the two portions (I and II) of the rectangle which do not lie inside our figure.

b. As in problem 148, choose bases of equal length for the n rectangles (fig. 88b). The sum S_n is given by

$$S_n = h(\sin h + \sin 2h + \cdots + \sin nh),$$

where

$$h = \frac{a}{n}.$$

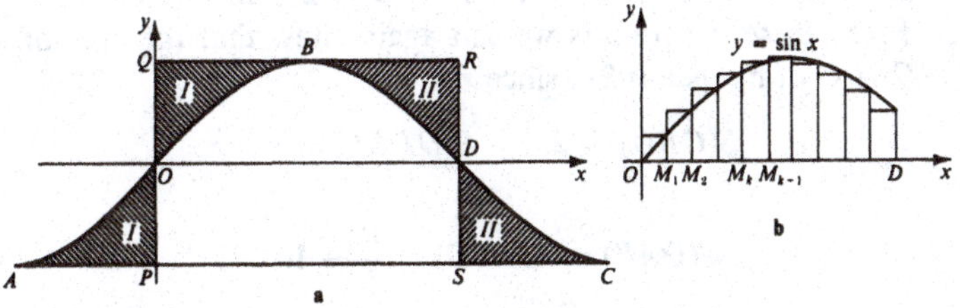

Fig. 88

Using the identity[10]

$$\sin h + \sin 2h + \cdots + \sin nh = \frac{\sin \frac{1}{2}nh \sin \frac{1}{2}(n+1)h}{\sin \frac{1}{2}h},$$

[10] This formula can be easily derived as follows: since

$$\sin\alpha \sin\beta = \tfrac{1}{2}[\cos(\alpha-\beta) - \cos(\alpha+\beta)]$$

we have

$$\begin{aligned}(\sin h + \sin 2h + \cdots + \sin nh)\sin\frac{h}{2} &= \frac{1}{2}\left[\left(\cos\frac{h}{2} - \cos\frac{3h}{2}\right) + \left(\cos\frac{3h}{2} - \cos\frac{5h}{2}\right)\right. \\ &\quad \left. + \cdots + \left(\cos\frac{(2n-1)}{2}h - \cos\frac{(2n+1)h}{2}\right)\right] \\ &= \frac{1}{2}\left[\cos\frac{h}{2} - \cos\left(n+\frac{1}{2}\right)h\right],\end{aligned}$$

and so

$$\begin{aligned}\sin h + \sin 2h + \cdots + \sin nh &= \frac{\frac{1}{2}\left[\cos\frac{h}{2} - \cos\left(n+\frac{1}{2}\right)h\right]}{\sin\frac{1}{2}h} \\ &= \frac{\sin\frac{1}{2}nh \sin\frac{1}{2}(n+1)h}{\sin\frac{1}{2}h}.\end{aligned}$$

we deduce that

$$S_n = h\frac{\sin \frac{1}{2}nh \sin \frac{1}{2}(n+1)h}{\sin \frac{1}{2}h} = \frac{a}{n}\frac{\sin \frac{1}{2}a \sin ((n+1)a/2n)}{\sin (a/2n)}.$$

But since $\lim_{\alpha\to 0} (\sin \alpha)/\alpha = 1$, we have

$$\lim_{n\to\infty} \frac{a/n}{\sin (a/2n)} = \lim_{n\to\infty} 2\frac{a/2n}{\sin (a/2n)} = 2.$$

It follows that the required area is

$$S = \lim_{n\to\infty} S_n = 2 \lim_{n\to\infty} \sin\frac{a}{2} \sin\left[\frac{a}{2}\left(1+\frac{1}{n}\right)\right] = 2\sin^2\frac{a}{2}.$$

In particular, when $\alpha = \pi$ we have $S = 2$; and when $\alpha = \pi/2$, $S = 1$. From these two results we can again show that the area of the figure $AOBDC$ of fig. 88a is 2π, since

$$S(OBD) = 2, \qquad S(ODSP) = \pi,$$

and

$$S(OAP) = S(DCS) = \frac{\pi}{2} - 1.$$

150a. In this problem the most convenient way to divide up the segment AD of the x axis (where $OA = a$, $OD = b$) is not by points chosen at equal intervals, but by points chosen so that

$$\frac{OM_1}{a} = \frac{OM_2}{OM_1} = \frac{OM_3}{OM_2} = \cdots = \frac{b}{OM_{n-1}}$$

(fig. 89). Let us denote the common value of these ratios by q. Then

$$\frac{OM_1}{a} = \frac{OM_2}{OM_1} = \frac{OM_3}{OM_2} = \cdots = \frac{b}{OM_{n-1}} = q,$$

$$\frac{b}{a} = \frac{OM_1}{a}\frac{OM_2}{OM_1}\frac{OM_3}{OM_2}\cdots\frac{b}{OM_{n-1}} = q^n,$$

so that $q = \sqrt[n]{b/a}$. Consequently, as $n \to \infty$, $q \to 1$.[11]

It is clear that

$$OM_1 = aq,$$

$$OM_2 = OM_1 \cdot q = aq^2,$$

$$OM_3 = OM_2 \cdot q = aq^3, \ldots, b = OM_{n-1} \cdot q = aq^n.$$

[11] Since q decreases as n increases, and $q \geqq 1$, it follows that $L = \lim_{n\to\infty} q$ exists, and $L \geqq 1$. Since $q \geqq L$, we have $b/a \geqq L^n$ for all n, which is impossible if $L > 1$. Hence $L = 1$.

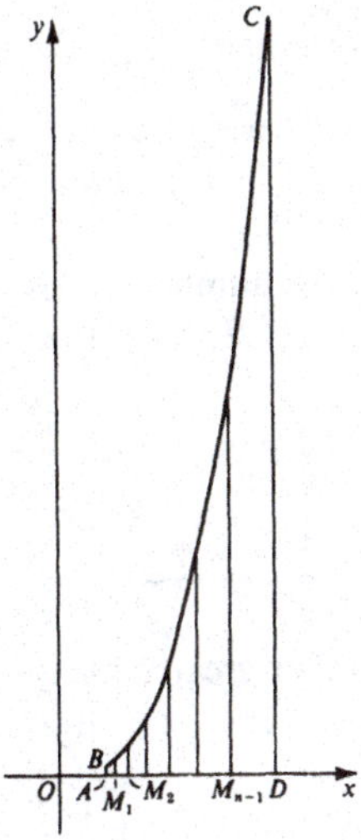

Fig. 89

Hence

$$h_1 = OM_1 - OA = a(q-1),$$

$$h_2 = aq^2 - aq = aq(q-1),$$

$$h_3 = aq^3 - aq^2 = aq^2(q-1),$$

$$\cdots\cdots\cdots\cdots\cdots\cdots\cdots\cdots$$

$$h_n = aq^n - aq^{n-1} = aq^{n-1}(q-1).$$

We note here that the necessary condition described on p. 26 is fulfilled: as $n \to \infty$, $q = \sqrt[n]{b/a} \to 1$, and the length h_k of each segment $M_{k-1}M_k$ tends to zero. The area of the kth rectangle is clearly

$$h_k(OM_k)^m = aq^{k-1}(q-1)(aq^k)^m,$$

so that the sum of the areas of all n rectangles is

$$\begin{aligned} S_n &= a(q-1)(aq)^m + aq(q-1)(aq^2)^m + aq^2(q-1)(aq^3)^m + \cdots \\ &\quad + aq^{n-1}(q-1)\cdot(aq^n)^m \\ &= a^{m+1}q^m(q-1)[1 + q^{m+1} + q^{2(m+1)} + \cdots + q^{(n-1)(m+1)}] \\ &= a^{m+1}q^m(q-1)\frac{q^{n(m+1)}-1}{q^{m+1}-1} \\ &= q^m \frac{1}{(q^{m+1}-1)/(q-1)} a^{m+1}\left[\left(\frac{b}{a}\right)^{m+1} - 1\right] \end{aligned}$$

[since $q^{n(m+1)} = (q^n)^{m+1} = (b/a)^{m+1}$]. As $n \to \infty, q \to 1$, so that $\lim_{n\to\infty} q^m = 1$ for any fixed m. Thus in order to determine the area of the curvilinear trapezoid $ABCD$ it suffices to find the limit of the expression $(q^{m+1} - 1)/(q - 1)$ as $q \to 1$.

We consider separately a number of cases.

(1) m is a positive integer. In this, the simplest case,

$$\lim_{q\to 1} \frac{q^{m+1} - 1}{q - 1} = \lim_{q\to 1} (q^m + q^{m-1} + \cdots + q + 1)$$
$$= \underbrace{1 + 1 + \cdots + 1 + 1}_{m+1 \text{ times}} = m + 1.$$

(2) m is a rational number greater than -1. We may write $m + 1 = r/s$, where r and s are positive integers. In this case

$$\lim_{q\to 1} \frac{q^{m+1} - 1}{q - 1} = \lim_{q\to 1} \frac{q^{r/s} - 1}{q - 1} = \lim_{q\to 1} \left[\frac{(q^{1/s})^r - 1}{q^{1/s} - 1} \div \frac{(q^{1/s})^s - 1}{q^{1/s} - 1}\right].$$

Now as $q \to 1$, $q_1 = \sqrt[s]{q} \to 1$. It therefore follows from the result of case (1) that

$$\lim_{q\to 1} \frac{(q^{1/s})^r - 1}{q^{1/s} - 1} = \lim_{q_1\to 1} \frac{q_1^{\,r} - 1}{q_1 - 1} = r, \qquad \lim_{q\to 1} \frac{(q^{1/s})^s - 1}{q^{1/s} - 1} = s,$$

and hence

$$\lim_{q\to 1} \frac{q^{m+1} - 1}{q - 1} = \frac{r}{s} = m + 1.$$

(3) m is rational and less than -1. Say $m + 1 = -r/s$, where r and s are positive integers. In this case, using the result of case (2), we find

$$\lim_{q\to 1} \frac{q^{m+1} - 1}{q - 1} = \lim_{q\to 1} \frac{q^{-r/s} - 1}{q - 1} = \lim_{q\to 1} \left(-q^{-r/s} \frac{q^{r/s} - 1}{q - 1}\right)$$
$$= -1 \cdot \frac{r}{s} = m + 1.$$

(4) If m is irrational, then q^{m+1} is defined as the limit of $q^{r/s}$, where r/s ranges through a set of rationals converging to $m + 1$. Since for any rational number $m \neq 1$, $\lim_{q\to 1} (q^{m+1} - 1)/(q - 1) = m + 1$, it follows[12] that

[12] To prove this, suppose for definiteness that $q > 1$. Let a positive number ε be given, and choose rational numbers r_1, r_2 such that

$$r_1 < m + 1 < r_2, \quad r_1 > m + 1 - \varepsilon, \quad \text{and} \quad r_2 < m + 1 + \varepsilon.$$

Then

$$\frac{q^{r_1} - 1}{q - 1} < \frac{q^{m+1} - 1}{q - 1} < \frac{q^{r_2} - 1}{q - 1},$$

for irrational m as well

$$\lim_{q\to 1}\frac{q^{m+1}-1}{q-1}=m+1.$$

Thus we see that for any $m \neq -1$, $\lim\limits_{q\to 1}(q^{m+1}-1)/(q-1)=m+1$. It follows that for any such m the area of the curvilinear trapezoid $ABCD$ of fig. 89 is

$$1\cdot\frac{1}{m+1}a^{m+1}\left[\left(\frac{b}{a}\right)^{m+1}-1\right]=\frac{b^{m+1}-a^{m+1}}{m+1}.$$

Remark. For $m = -1$ the formula for the area of $ABCD$ is of an entirely different form; see problem 154 below.

b. To calculate the area of the curvilinear triangle bounded by the curve $y = x^m$, the x axis, and the line $x = b$, the method we used for the previous question is inapplicable. This is because $OD = a$ cannot be divided into a finite number of segments in such a way that the coordinates

---

since $q > 1$. We have already shown that

$$\lim_{q\to 1}\frac{q^{r_1}-1}{q-1}=r_1$$

and

$$\lim_{q\to 1}\frac{q^{r_2}-1}{q-1}=r_2.$$

Hence, if q is sufficiently close to 1, we have

$$\frac{q^{r_1}-1}{q-1}>r_1-\varepsilon,\qquad \frac{q^{r_2}-1}{q-1}<r_2+\varepsilon.$$

Then

$$r_1-\varepsilon<\frac{q^{m+1}-1}{q-1}<r_2+\varepsilon.$$

Since $r_1 > m+1-\varepsilon$ and $r_2 < m+1+\varepsilon$, it follows that

$$m+1-2\varepsilon<\frac{q^{m+1}-1}{q-1}<m+1+2\varepsilon.$$

(The same inequality is obtained for $q < 1$ by choosing $r_1 > m+1 > r_2$, $r_1 < m+1+\varepsilon$, $r_2 > m+1-\varepsilon$.) Thus, if $\varepsilon > 0$ is given, we have

$$\left|\frac{q^{m+1}-1}{q-1}-(m+1)\right|<2\varepsilon$$

for all q sufficiently close to 1. This means by definition that

$$\lim_{q\to 1}\frac{q^{m+1}-1}{q-1}=m+1.$$

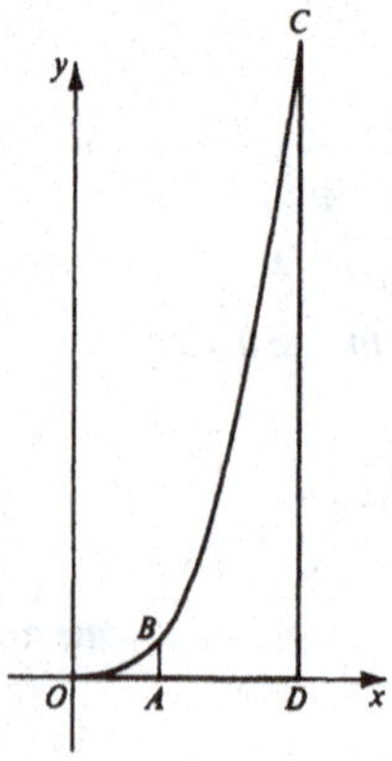

Fig. 90

of the endpoints form a geometric progression. However, we can use the result of the previous problem. The area of the curvilinear triangle OCD (fig. 90) is the limit as $a \to 0$ of the area of the curvilinear trapezoid $ABCD$. The expression $(b^{m+1} - a^{m+1})/(m + 1)$ tends to $b^{m+1}/(m + 1)$ as a tends to zero; it follows that the area of OCD is $b^{m+1}/(m + 1)$.

Remark. The result of part **b** remains valid even for negative values of m, provided they are greater than -1. For if $m > -1$ then $\lim_{a \to 0} (b^{m+1} - a^{m+1})/(m+1) = b^{m+1}/(m + 1)$. In case $m < 0$ the curvilinear triangle OCD no longer exists, since the curve $y = x^m$ has the form shown in fig. 91a. (The reader should examine for himself the intermediate case $m = 0$, and compare its graph with figs. 90 and 91.) The figure bounded by the curve, the x axis, the line $x = 0$, and $x = b$ is unbounded for $m < 0$, but we may nevertheless speak of its area. We may, for example, define the area to be the limit of the sum of the areas of the rectangles shown in 91a as the length of the base of each of them tends to zero; this limit is equal to $b^{m+1}/(m + 1)$.

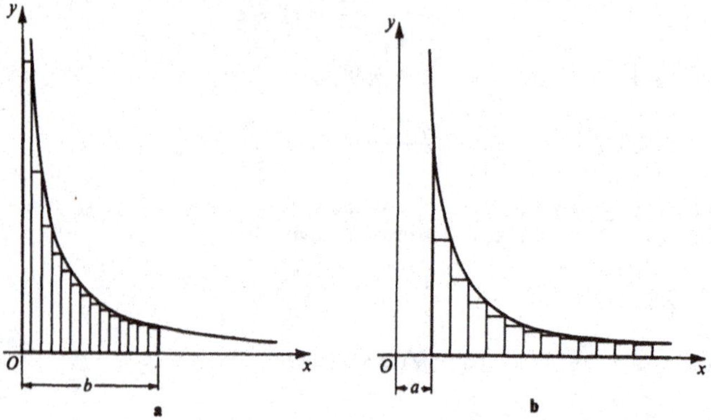

Fig. 91

In an analogous manner, when $m < -1$ we can speak of the area of the figure bounded by the curve $y = x^m$, the x axis, and the line $x = a$, although this figure is unbounded (fig. 91b). This area may be defined as the limit of the sum of the areas of the rectangles shown in fig. 91b as the number of rectangles tends to infinity and the length of the base of each rectangle tends to zero. The area so defined is equal to

$$\lim_{b\to\infty} \frac{b^{m+1} - a^{m+1}}{m+1} = \frac{a^{m+1}}{-(m+1)}.$$

151. Suppose $OA_1 = a_1$, $OD_1 = b_1$ and $OA_2 = a_2$, $OD_2 = b_2$. Divide the segments $A_1D_1A_2D_2$ into n equal parts and construct a rectangle on each part as base, as shown in fig. 92. Clearly the area of the kth rectangle

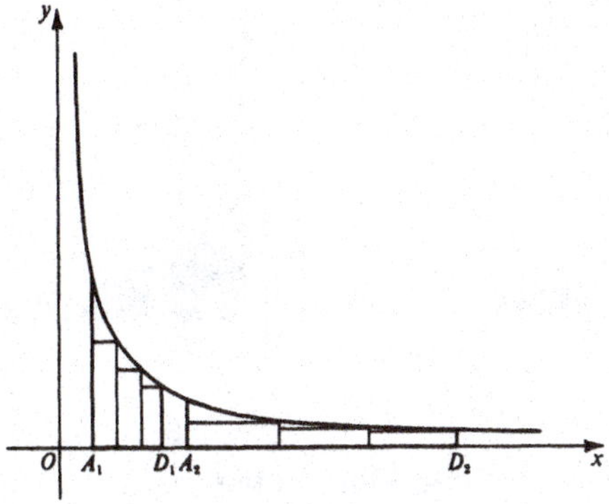

Fig. 92

of the first set is

$$A_k^{(1)} = \frac{b_1 - a_1}{n} \frac{1}{a_1 + k(b_1 - a_1)/n}$$

$$= \frac{b_1 - a_1}{(n-k)a_1 + kb_1} = \frac{b_1/a_1 - 1}{n - k + k(b_1/a_1)}.$$

We can find the area of the kth rectangle of the second set in the same way; it is

$$A_k^{(2)} = \frac{b_2 - a_2}{n} \frac{1}{a_2 + k(b_2 - a_2)/n}$$

$$= \frac{b_2 - a_2}{(n-k)a_2 + kb_2} = \frac{b_2/a_2 - 1}{n - k + k(b_2/a_2)}.$$

Since $b_2/a_2 = b_1/a_1$, we have $A_k^{(1)} = A_k^{(2)}$; it follows that the sum $S_n^{(1)}$ of the areas of the first n rectangles is equal to the sum $S_n^{(2)}$ of the areas of the second n rectangles. Since n is arbitrary, the required result follows from the definition of the area under a curve (see p. 26).

152. We consider separately a number of cases.

(1) z_1 and z_2 are both greater than 1 (fig. 93a). Since $(z_1z_2)/z_2 = z_1/1$ it follows from problem 151 that the areas of the two shaded trapezoids in fig. 93a are equal. Since the area of the trapezoid bounded by the x axis, the hyperbola $y = 1/x$, and the lines $x = 1$ and $x = z_1z_2$ is equal to the sum of the areas of the two trapezoids bounded by the x axis and the curve, and the lines $x = 1$, $x = z_2$ and $x = z_2$ and $x = z_1z_2$, respectively,

$$F(z_1z_2) = F(z_1) + F(z_2).$$

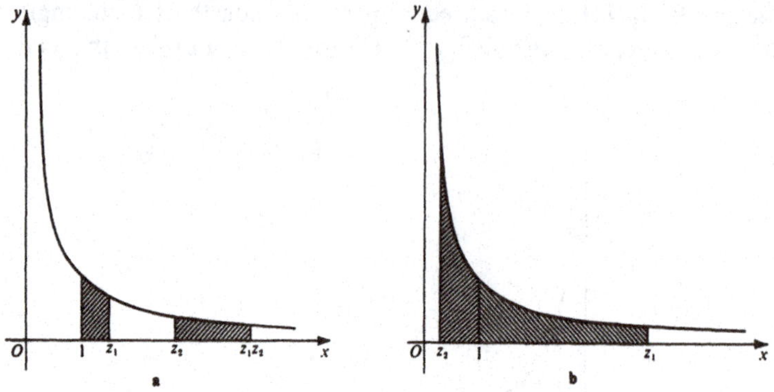

Fig. 93

(2) $z_2 = 1/z_1$, $z_1 > 1$ (fig. 93b). In this case the identity to be proved assumes the form

$$F(z_1) + F\left(\frac{1}{z_1}\right) = F(1) = 0$$

or

$$F\left(\frac{1}{z_1}\right) = -F(z_1).$$

Since $z_1{:}1 = 1{:}1/z_1$, it follows from problem 151 that the areas of the trapezoids bounded by the curve, the x axis, and the pairs of lines $x = 1$, $x = z_1$, and $x = 1/z_1$, $x = 1$, respectively, have the same area. The equation $F(1/z_1) = -F(z_1)$ follows immediately from this.

(3) z_1 and z_2 are both less than 1. In this case $1/z_1$, $1/z_2$, $1/z_1z_2$ are all greater than 1, and from what we have already proved

$$F\left(\frac{1}{z_1}\right) = -F(z_1), \quad F\left(\frac{1}{z_2}\right) = -F(z_3), \quad F\left(\frac{1}{z_1z_2}\right) = -F(z_1z_2),$$

and

$$F\left(\frac{1}{z_1}\right) + F\left(\frac{1}{z_2}\right) = F\left(\frac{1}{z_1z_2}\right).$$

Multiplying both sides of this equality by -1, we arrive at the required identity

$$F(z_1) + F(z_2) = F(z_1 z_2).$$

(4) $z_1 > 1$, $z_2 < 1$ but $z_1 z_2 \neq 1$. Suppose for definiteness that $z_1 > 1/z_2$, so that

$$z_1 z_2 > 1.$$

Since $1/z_2 > 1$ we may use the results already established to write

$$F(z_1 z_2) + F\left(\frac{1}{z_2}\right) = F\left(z_1 z_2 \cdot \frac{1}{z_2}\right) = F(z_1),$$

whence

$$F(z_1 z_2) = F(z_1) - F\left(\frac{1}{z_2}\right) = F(z_1) + F(z_2).$$

The case where $z_1 z_2 < 1$ is dealt with similarly.

We have thus solved the problem in all cases.

153. Note that by its definition the function F is *increasing*, that is, for $z_2 > z_1$, $F(z_2) > F(z_1)$. Since $F(1) = 0$ and F increases continuously as z increases (and so cannot "skip" any values), in order to prove the existence of a real number e such that $F(e) = 1$, it is sufficient to show that there exist values z for which $F(z) > 1$.

We will show that $F(3) > 1$. Draw a tangent $B'C'$ to the hyperbola at the point $G(2,\frac{1}{2})$ (fig. 94a). The area of the trapezoid $ABC'D$ (where $OA = 1$, $OD = 3$) is 1, since its midline $HG = \frac{1}{2}$, while its height AD is 2. It follows that the area $F(3)$ of the curvilinear triangle $ABCD$ is greater than 1, as required.

We note now that $F(2) < 1$. This follows from fig. 94b: the area of the curvilinear triangle $ABGH$ ($OA = 1$; $OH = 2$) is equal to $F(2)$, and is clearly less than the area of the square $ABG'H$, which is 1.

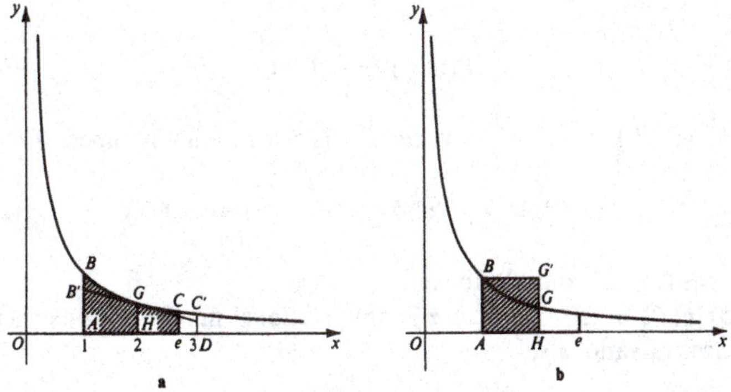

Fig. 94

We thus conclude that

$$F(2) < 1 < F(3)$$

and so

$$2 < e < 3.$$

154. We first prove the following important property of the function F: for any α,

$$F(z^\alpha) = \alpha F(z).$$

We give the proof in a number of steps, each time extending the range of α for which the theorem holds.

(1) $F(z^n) = nF(z)$ when n is a positive integer. For by the result of problem 152,

$$F(z^2) = F(z \cdot z) = F(z) + F(z) = 2F(z),$$

$$F(z^3) = F(z^2 \cdot z) = F(z^2) + F(z) = 2F(z) + F(z) = 3F(z),$$

$$F(z^4) = F(z^3 \cdot z) = F(z^3) + F(z) = 3F(z) + F(z) = 4F(z),$$

$$\vdots$$

$$F(z^n) = F(z^{n-1} \cdot z) = F(z^{n-1}) + F(z) = (n-1)F(z) + F(z) = nF(z).$$

(2) $F(z^k) = kF(z)$ when k is a negative integer. For let $k = -n$, where n is a positive integer. Then clearly $F(z^k) = F(1/z^n) = -F(z^n)$ (see the solution to problem 142), and therefore

$$F(z^k) = -nF(z) = kF(z).$$

(3) $F(z^{1/m}) = (1/m)F(z)$ for m an integer. For let us substitute $z^{1/m}$ for z_1 in the identity $F(z_1{}^m) = mF(z_1)$. Then we obtain

$$F(z) = mF(z^{1/m}),$$

whence

$$F(z^{1/m}) = \frac{1}{m} F(z).$$

(4) $F(z^{n/m}) = n/m\, F(z)$, where n/m is any rational number. For

$$F(z^{n/m}) = F[(z^{1/m})^n] = nF(z^{1/m}) = \frac{n}{m} F(z).$$

[See cases (1), (2), and (3) above.]

(5) $F(z^\alpha) = \alpha F(z)$, where α is any irrational number. Construct two sequences of rationals,

$$\alpha_1, \alpha_2, \ldots, \alpha_n, \ldots \quad \text{and} \quad \alpha_1', \alpha_2', \ldots, \alpha_n', \ldots$$

satisfying the following conditions:

$$\alpha_n < \alpha < \alpha_n', \qquad \lim_{n\to\infty} \alpha_n = \lim_{n\to\infty} \alpha_n' = \alpha.$$

(We could, for example, choose α_n and α_n' as follows. Let the representation of α in the decimal system be $a_k a_{k-1} \cdots a_0 \cdot b_1 b_2 b_3 \ldots$. Put $\alpha_n = a_k a_{k-1} \cdots a_0 \cdot b_1 b_2 \cdots b_n$, and $\alpha_n' = \alpha_n + 10^{-n}$. Then $0 < \alpha - \alpha_n < 10^{-n}$ and $0 < \alpha_n' - \alpha_n < 10^{-n}$.)

Since F is clearly an increasing function, we have the inequalities

$$F(z^{\alpha_n}) < F(z^{\alpha}) < F(z^{\alpha_n'}),$$

or, since α_n and α_n' are rationals,

$$\alpha_n F(z) < F(z^{\alpha}) < \alpha_n' F(z).$$

[See case (4)]. Thus

$$\alpha_n < \frac{F(z^{\alpha})}{F(z)} < \alpha_n',$$

and therefore

$$\lim_{n\to\infty} \alpha_n \leqq \frac{F(z^{\alpha})}{F(z)} \leqq \lim_{n\to\infty} \alpha_n'.$$

Since

$$\lim_{n\to\infty} \alpha_n = \alpha = \lim_{n\to\infty} \alpha_n',$$

it follows from the last inequality that

$$\frac{F(z^{\alpha})}{F(z)} = \alpha, \qquad F(z^{\alpha}) = \alpha F(z),$$

as required.

It is now easy to show that

$$F(z) = \log_e z.$$

For by the above

$$F(z) = F(e^{\log_e z}) = \log_e z F(e) = \log_e z,$$

since $F(e) = 1$ by the definition of e.

155. In this problem the same experiment is considered as in problem 96 of Volume I. As in the second solution to that problem, we can characterize all possible outcomes to this experiment in terms of the two numbers $AK = x$ and $KL = z$ (see fig. 95). Then the set of all possible outcomes to the experiment can be represented as the set of all points in or on the

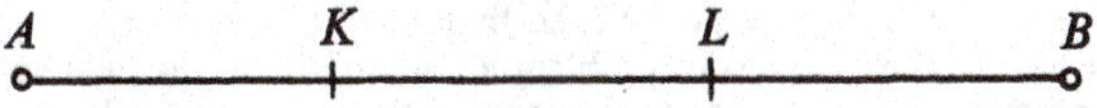

Fig. 95

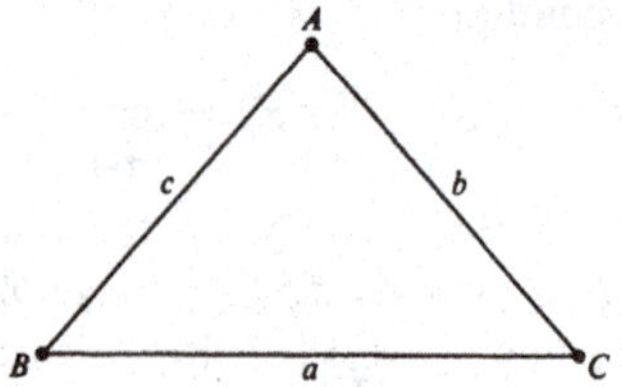

Fig. 96

triangle OST bounded by the coordinate axes and the line $x + z = l$ (fig. 97). The probability of any event is the ratio of the area of the part of OST corresponding to the points at which the event occurs to the area of the entire triangle OST.

We must now determine for which values of x and z an acute-angled triangle can be formed from three segments with lengths x, z, and $l - x - z$. A necessary and sufficient condition is that the square of each of these lengths be less than the sum of the squares of the other two lengths.[13] Therefore the unfavorable outcomes of the experiment correspond to the points of the triangle OST for which at least one of the following inequalities is fulfilled:

$$x^2 \geqq z^2 + (l - x - z)^2, \qquad z^2 \geqq x^2 + (l - x - z)^2, \qquad (l - x - z)^2 \geqq x^2 + z^2. \tag{1}$$

Moreover, no two of the above inequalities can be satisfied simultaneously (for if r, s, t are three numbers with $r \leqq s \leqq t$, then $r^2 \leqq s^2 + t^2$ and $s^2 \leqq r^2 + t^2$). Consequently, no two of the three regions determined by these inequalities overlap. Furthermore, in splitting the rod at random into three pieces, the probability that the square of the length of a given piece is greater than or equal to the sum of the squares of the lengths of the other two pieces is the same for each of the three pieces of the rod[14]). It follows from this that the areas of the three regions of the triangle OST determined by the three inequalities (1) are the same. Hence in order to find the area of the region corresponding to all the unfavorable outcomes of the experiment, we need only compute the area of the region defined by one of the three inequalities (1) and multiply it by 3.

[13] The necessity of this condition follows from the law of cosines. For if a, b, c are the sides of an acute triangle (fig. 96), then $a^2 = b^2 + c^2 - 2bc \cos A < b^2 + c^2$, since $\cos A > 0$. Similarly, $b^2 < a^2 + c^2$ and $c^2 < a^2 + b^2$. Conversely, if a, b, c are any three positive numbers satisfying $a^2 < b^2 + c^2$, $b^2 < a^2 + c^2$, $c^2 < a^2 + b^2$, then $a^2 < b^2 + 2bc + c^2 = (b + c)^2$, so that $a < b + c$. Similarly, $b < a + c$ and $c < a + b$. Hence there is a triangle ABC with sides a, b, c. From the law of cosines for ABC we find that $\cos A$, $\cos B$, $\cos C$ are positive, so that A, B, C are acute.

[14] As we saw in the solution to problem 98 of Volume I, the problem of breaking a rod into three pieces is equivalent to the problem of dividing a circle into three pieces, and in the division of the circle the three pieces obtained are obviously interchangeable.

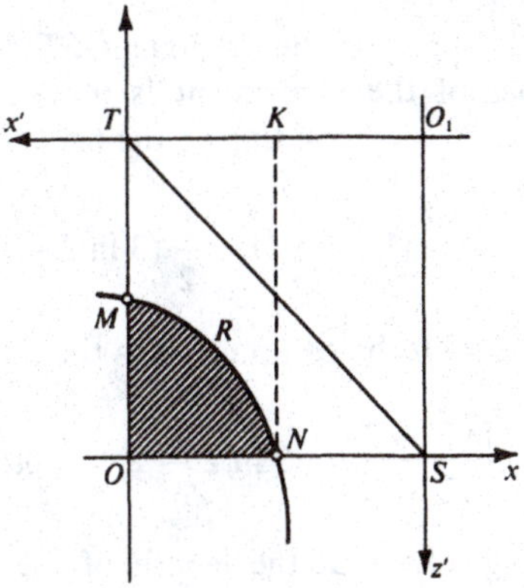

Fig. 97

Let us now find the area of the part of the triangle OST for which the inequality

$$(l - x - z)^2 \geqq x^2 + z^2$$

is satisfied. Removing parentheses, this inequality can be rewritten in the form

$$l^2 + 2xz - 2lx - 2lz \geqq 0.$$

Adding l^2 to each side of the inequality and dividing by 2, we obtain:

$$(l - x)(l - z) \geqq \frac{l^2}{2}.$$

But the equation $(l - x)(l - z) = l^2/2$ defines a hyperbola which passes through the points $x = \frac{1}{2}l$, $z = 0$ (the point N) and $x = 0$, $z = \frac{1}{2}l$ (the point M); the inequality $(l - x)(l - z) \geqq l^2/2$ is satisfied by all points of the curved region $ONRM$ (the shaded part of fig. 97) and only by those points.

It remains for us to determine the area of the region $ONRM$. Introduce a new coordinate system by taking for coordinate axes the lines O_1T and O_1S (that is, the lines $l - z = 0$ and $l - x = 0$; see fig. 97); in this coordinate system our hyperbola will have the equation $x'z' = l^2/2$. Now take as a new unit of length $l/\sqrt{2}$. Then the abscissa O_1K of the point N in the new units is $(l/2)/(l/\sqrt{2}) = \sqrt{2}/2$, and the abscissa O_1T of the point M is $l/(l/\sqrt{2}) = \sqrt{2}$. By the result of problem 154, the area of the hyperbolic trapezoid $KNMT$ in the new units is $\ln(\sqrt{2}/(\sqrt{2}/2)) = \ln 2$. Passing back to the old unit of area, we find that the area of the hyperbolic trapezoid $KNMT$ is $(\ln 2)(l^2/2) = (l^2 \ln 2)/2$. Hence for the area of the shaded region $ONRM$ of fig. 97, we obtain

$$\frac{l^2}{2} - \frac{l^2 \ln 2}{2} = \frac{l^2}{2}(1 - \ln 2).$$

The total area of the part of the triangle OST corresponding to the unfavorable outcomes of the experiment is three times this expression, and the area of the part corresponding to the favorable outcomes is

$$\frac{l^2}{2} - 3\frac{l^2}{2}(1 - \ln 2) = \frac{l^2}{2}(3\ln 2 - 2).$$

Therefore the probability to be computed has the following value:

$$\frac{(l^2/2)(3\ln 2 - 2)}{l^2/2} = 3\ln 2 - 2 = 0.082\cdots$$

156. Suppose for simplicity that the length of the rod is 2 (this is no restriction, since we can always take half the length of the rod as the unit

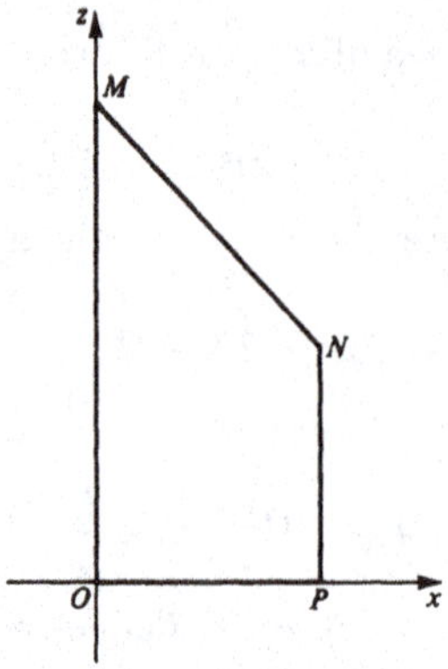

Fig. 98

of length). Let AB be the rod, and K and L the first and second break points. The outcome of the experiment in this problem is completely determined by the numbers $x = AK$ and $z = KL$, where x is chosen at random from the interval $0 \leqq x \leqq 1$ (x is the length of the smaller of the two segments formed after breaking the rod the first time, whence $x \leq 1$) and z is chosen at random from the interval $0 \leqq z \leqq 2 - x$. If we take x and z as coordinates in the plane, then the set of all possible outcomes to the experiment is represented by the set of all points on or inside the trapezoid $OMNP$ (fig. 98). However, the probability that (x,z) falls within some small rectangle located within the trapezoid does not depend only on the area of this rectangle: the probability that x falls within a given segment of the x axis is the length of that segment, but the probability that z falls within any segment of the z axis for given x is the ratio of the length of that segment to $2 - x$, and hence depends on x. Therefore the probability that a point (x,z) falls within any part of the trapezoid $OMNP$ depends not only on the area of that part but also on its location within $OMNP$; accordingly, the representation of all possible outcomes of the experiment as the points on or in the trapezoid $OMNP$ is inconvenient for computing probabilities.

It is more convenient to characterize all possible outcomes to the experiment in terms of the numbers x and $y = z/(2 - x)$. The selection of the two break points amounts to a random selection of x and y from between 0 and 1; consequently, if we take x and y as coordinates in the plane, then the possible outcomes to the experiment are represented by the points of the unit square $OUVW$ (fig. 99), and the probability that a point

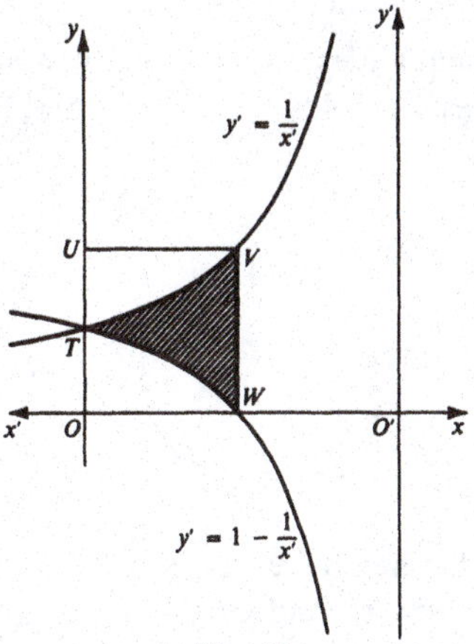

Fig. 99

(x,y) falls within some part of this square is the area of that part (since the area of the entire square $OUVW$ is 1). Hence we need only determine the area of the part of the square $OUVW$ which corresponds to the favorable outcomes of the experiment.

In this problem the favorable outcomes are those in which a triangle can be formed from segments of lengths x, z, and $2 - x - z$. In order that this be possible, it is necessary and sufficient that none of these lengths exceed 1. Since x is by hypothesis at most 1, the favorable outcomes are those in which $z < 1$ and $2 - x - z < 1$, that is,

$$z < 1 \quad \text{and} \quad z > 1 - x.$$

Since $y = z/(2 - x)$, we find by dividing the above inequalities by $2 - x$ that the favorable outcomes correspond to the points (x,y) for which $y < 1/(2 - x)$ and $y > (1 - x)/(2 - x)$, that is,

$$y < \frac{1}{2 - x}, \qquad y > 1 - \frac{1}{2 - x}. \tag{1}$$

We now have to determine what the region in fig. 99 defined by these inequalities looks like.

In place of the coordinates (x,y) let us introduce coordinates (x',y'), where

$$y = y' \quad \text{and} \quad x' = 2 - x.$$

This change of coordinates amounts to a translation of the origin to the point O' located on the x axis at a distance of 2 from O, and a reversal of the direction of the x axis (see fig. 99). In the new coordinates the inequalities (1) assume the form:

$$y' < \frac{1}{x'}, \qquad y' > 1 - \frac{1}{x'}. \tag{2}$$

The first of these inequalities shows that all points which correspond to favorable outcomes lie below the hyperbola

$$y' = \frac{1}{x'}.$$

Now consider the curve

$$y' = 1 - \frac{1}{x'} = \frac{1}{2} - \left(\frac{1}{x'} - \frac{1}{2}\right);$$

$$y' - \frac{1}{2} = -\left(\frac{1}{x'} - \frac{1}{2}\right);$$

this curve is the one obtained by reflecting the curve $y' = 1/x'$ over the line $y' = \frac{1}{2}$ (see fig. 99). The second of the inequalities (2) shows that all points which correspond to favorable outcomes are located below the curve

$$y' = 1 - \frac{1}{x'}.$$

Therefore the favorable outcomes correspond to the points of the shaded region in fig. 99. We must determine the area of this region.

Note first that by virtue of the symmetry of the curves

$$y' = \frac{1}{x'} \quad \text{and} \quad y' = 1 - \frac{1}{x'}$$

about the line $y' = \frac{1}{2}$, the areas of the hyperbolic triangles OTW and TUV are equal. Therefore the required area is $1 - 2$ area (TUV). But by problem 154, the area of the hyperbolic trapezoid $OTVW$ is $\ln 2$. It follows from this that area $TUV = 1 - \ln 2$, and, consequently, the area of the hyperbolic triangle TVW is

$$1 - 2(1 - \ln 2) = 2 \ln 2 - 1 = 0.388 \ldots.$$

This expression is the probability that a triangle can be formed from the segments x, z, and $2 - x - z$.

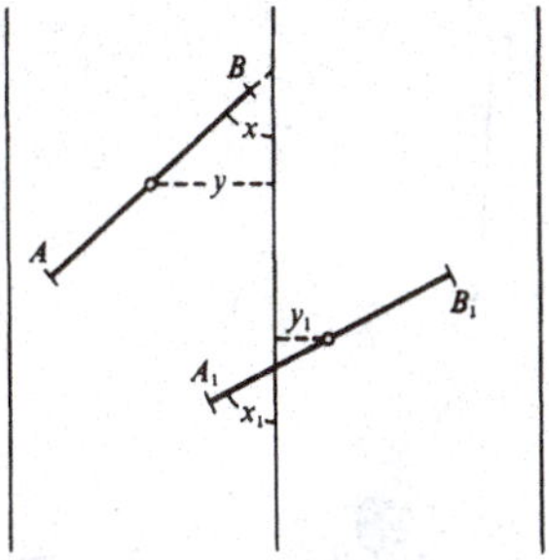

Fig. 100

157. The location of a needle AB after it lands on the plane is determined by the distance y from its center to the nearest of the parallel rulings and the angle x between AB and the rulings (fig. 100). It is clear that $0 \leqq y \leqq a$, $0 \leqq x \leqq \pi$; the possible outcomes to the experiment are determined by pairs of numbers (x,y), where x is chosen at random from between 0 and π, and y is chosen from between 0 and a. The set of all outcomes can be represented as the set of all points in or on a rectangle $OMNP$ with sides $OM = a$ and $OP = \pi$ (fig. 101). The probability of the point (x,y) lying within any region of this rectangle is the ratio of the area of that region to the area of the entire rectangle $OMNP$. We therefore need only determine the area of the part of $OMNP$ which corresponds to the favorable outcomes of the experiment.

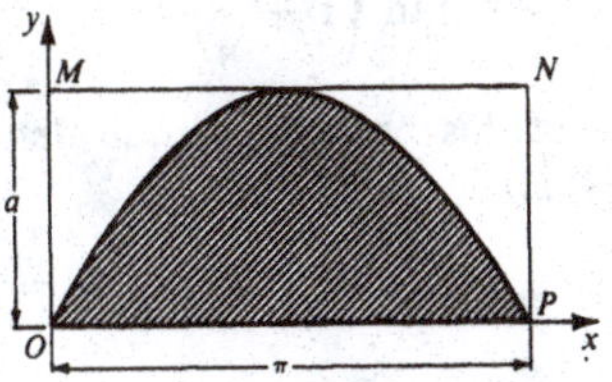

Fig. 101

It is apparent from fig. 100 that in order for the needle AB to lie across one of the rulings, it is necessary and sufficient that $y \leqq a \sin x$. Hence the favorable outcomes to the experiment correspond to the points located under this sinusoid. By problem 149**b**, the area under the sinusoid is $2a$. Since the area of the entire rectangle $OMNP$ is $a\pi$, the required probability is

$$\frac{2a}{a\pi} = \frac{2}{\pi} \approx 0.637.$$

158a. We know that the quantity $\ln(1 + 1/n)$ is equal to the area of the curvilinear trapezoid $ABCD$ bounded by the hyperbola $y = 1/x$, the x axis, and the lines $x = 1$ and $x = 1 + 1/n$ (see fig. 102). Since the area of

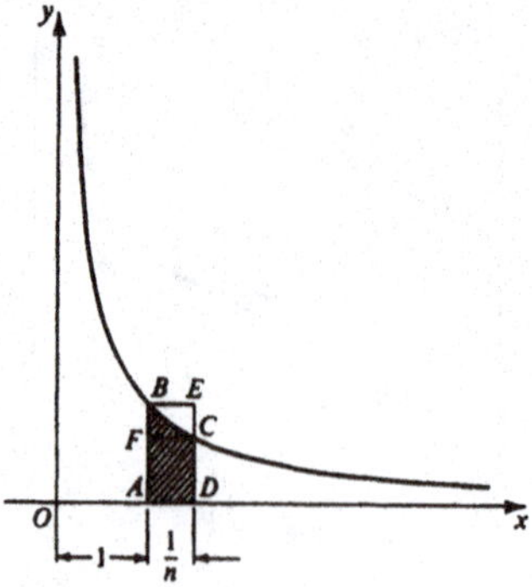

Fig. 102

this trapezoid is less than the area of the rectangle $ABED$ and greater than that of the rectangle $AFCD$,

$$\frac{1}{n} > \ln\left(1 + \frac{1}{n}\right) > \frac{1}{n} \cdot \frac{1}{1 + \frac{1}{n}} = \frac{1}{n+1}.$$

Hence

$$1 > n \ln\left(1 + \frac{1}{n}\right) > \frac{n}{n+1} = 1 - \frac{1}{n+1},$$

and therefore

$$\lim_{n\to\infty} n \ln\left(1 + \frac{1}{n}\right) = 1.$$

Remark. The result of this problem is equivalent to $\lim_{n\to\infty}(1 + 1/n)^n = e$. (See the end of problem 162.)

b. We know that $\log_a x = (\log_e x)/(\log_e a) = (\ln x)/\ln a$. (See footnote on p. 159.) Hence

$$\lim_{n\to\infty} n \log_a\left(1 + \frac{1}{n}\right) = \lim_{n\to\infty} \frac{n \ln(1 + 1/n)}{\ln a} = \frac{1}{\ln a}.$$

c. Recall that the area of the curvilinear trapezoid bounded by the hyperbola $y = 1/x$, the x axis, and the lines $x = 1$ and $x = \sqrt[n]{a}$ is equal to

$$\ln \sqrt[n]{a} = \frac{1}{n} \ln a.$$

The area of the trapezoid with sides $x = 1$ and $x = 1 + (1/n) \ln a$ is clearly less than the area of the rectangle with base $(1/n) \ln a$ and height 1 (see fig. 103a), that is, less than $(1/n) \ln a$. It follows that

$$\sqrt[n]{a} > 1 + \frac{1}{n} \ln a.$$

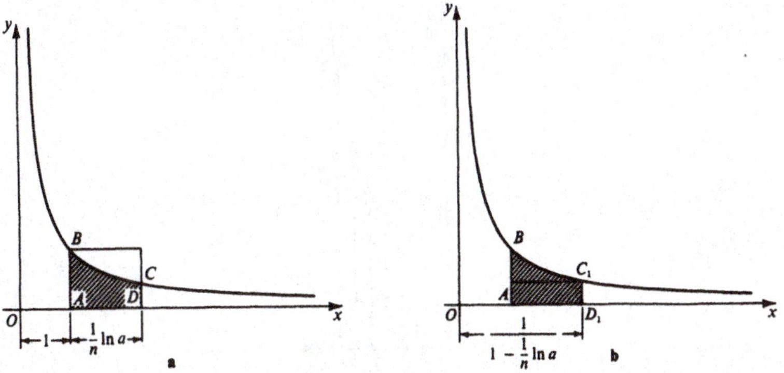

Fig. 103

In the same way we can show that the area of the trapezoid with sides $x = 1$ and $x = \dfrac{1}{1 - (1/n)\ln a}$ is greater than the area of the rectangle with base $\dfrac{1}{1 - (1/n)\ln a} - 1$ and height $1 - (1/n)\ln a$ (fig. 103b), that is, greater than $(1/n)\ln a$. It follows that

$$\sqrt[n]{a} < \frac{1}{1 - (1/n)\ln a}.$$

Thus

$$\frac{1}{n}\ln a < \sqrt[n]{a} - 1 < \frac{1}{1 - (1/n)\ln a} - 1 = \frac{(1/n)\ln a}{1 - (1/n)\ln a},$$

that is,

$$\ln a < n(\sqrt[n]{a} - 1) < \frac{\ln a}{1 - (1/n)\ln a},$$

Hence

$$\lim_{n\to\infty} n(\sqrt[n]{a} - 1) = \ln a.$$

159a. Divide the segment OD of the x axis, where $OD = b$, into n equal parts (fig. 104). The length of each part is b/n, and the heights of the rectangles having these parts as bases and inscribed in the curve $y = a^x$ are equal to $1, a^{b/n}, a^{2b/n}, \ldots, a^{(n-1)b/n}$. It follows that the total area of all these rectangles is

$$S_n = \frac{b}{n}(1 + a^{b/n} + a^{2b/n} + \cdots + a^{(n-1)b/n})$$
$$= \frac{b}{n}\frac{a^b - 1}{a^{b/n} - 1}.$$

The area S of the curvilinear trapezoid $OBCD$ is the limit of S_n as $n \to \infty$. Thus

$$S = \lim_{n\to\infty} \frac{b}{n}\cdot\frac{(a^b - 1)}{(a^{b/n} - 1)} = \lim_{n\to\infty} \frac{a^b - 1}{(n/b)(a^{b/n} - 1)}.$$

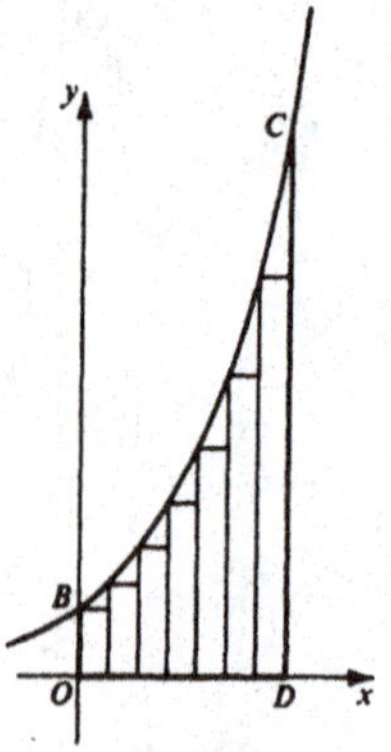

Fig. 104

But by the result of problem 158c,

$$\lim_{n\to\infty} \frac{n}{b}(a^{n/b} - 1) = \frac{1}{b}\lim_{n\to\infty} n(\sqrt[n]{a^b} - 1) = \frac{1}{b}\ln a^b = \ln a.$$

Thus

$$S = \frac{a^b - 1}{\ln a}.$$

(In particular, if $a = e$ we obtain the simple expression $S = e^b - 1$.)

b. *First solution.* We use part **a** and the fact that the equation $y = \log_a x$ can also be written in the form $x = a^y$. We must find the area of the curvilinear triangle ABC, where $OA = 1$ and $OB = b$ (fig. 105).

We first calculate the area of the curvilinear trapezoid $OACD$. The equation of our curve is $y = \log_a x$ or $x = a^y$. It follows that the curvilinear trapezoid $OACD$ is congruent to the curvilinear trapezoid $OBCD$ in fig. 104, whose area we calculated in the previous problem.[15] The trapezoid $OACD$ is bounded by the curve $x = a^y$ and the straight lines

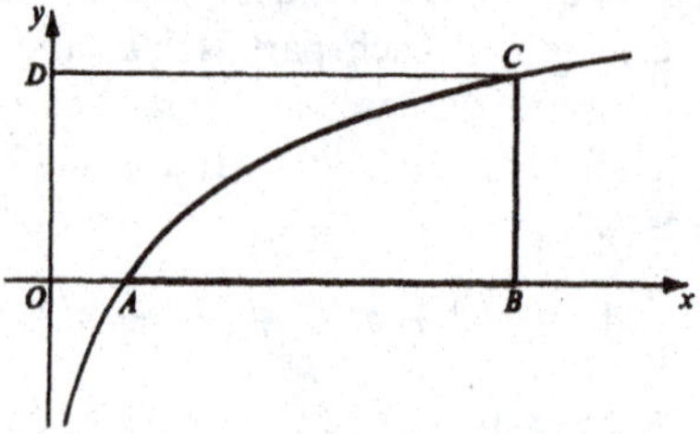

Fig. 105

[15] More precisely, these trapezoids are symmetric (since looking toward the positive direction of the x axis, the positive direction of the y axis is to the left, whereas looking toward the positive direction of the y axis, the positive direction of the x axis is to the right).

$y = 0$ and $y = \log_a b$. It follows from the result of part **a** that its area is

$$S(OACD) = \frac{a^{\log_a b} - 1}{\ln a} = \frac{b - 1}{\ln a}.$$

But the area of the curvilinear triangle can be found by subtracting this area from that of $OBCD$. We obtain[16]

$$S = b \log_a b - \frac{b - 1}{\ln a} = \frac{b \log_a b \ln a - b + 1}{\ln a} = \frac{b \ln b - b + 1}{\ln a}.$$

In particular, the area S_0 of the curvilinear triangle bounded by the x axis, the curve $y = \ln x$, and the straight line $x = b$ is given by

$$S_0 = b \ln b - b + 1,$$

and the area S_1 of the curvilinear triangle bounded by the x axis, the curve $y = \log x$, and the line $x = b$ is given by

$$S_1 = \frac{b \ln b - b + 1}{\ln 10} = \frac{\ln 10 \cdot b \log b - b + 1}{\ln 10} \approx \frac{2.3b \log b - b + 1}{2.3}.$$

Second solution. Let $OA = 1$, $OB = b$. Divide AB into n parts by points $M_1, M_2, \ldots, M_{n-1}$ so that

$$\frac{OM_1}{1} = \frac{OM_2}{OM_1} = \frac{OM_3}{OM_2} = \cdots = \frac{b}{OM_{n-1}}.$$

We have

$$OM_1 = q, \qquad OM_2 = q^2,$$
$$OM_3 = q^3, \ldots, OM_{n-1} = q^{n-1}, \qquad OM_n = q^n = b,$$

where $q = \sqrt[n]{b}$ (compare the solution of problem 150). It follows from this that the lengths of the bases of the rectangles in fig. 106 are

$$(q - 1), q(q - 1), q^2(q - 1), \ldots, q^{n-1}(q - 1)$$

and the heights are

$$\log_a q, \log_a q^2 = 2 \log_a q, \log_a q^3 = 3 \log_a q, \ldots, \log_a q^n = n \log_a q.$$

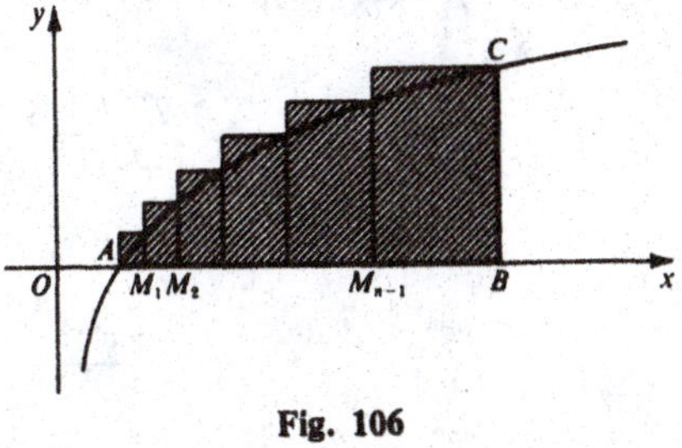

Fig. 106

[16] We use here the identity $\log_a b \ln a = \ln b$. It can be deduced from the following chain of identities:

$$e^{\ln b} = b = a^{\log_a b} = (e^{\ln a})^{\log_a b} = e^{\ln a \log_a b}.$$

Thus the area of the curvilinear triangle ABC is equal to the limit as $n \to \infty$ of the sum

$$S_n = (q-1)\log_a q + q(q-1)\cdot 2\log_a q + q^2(q-1)\cdot 3\log_a q + \cdots + q^{n-1}(q-1)\cdot n\log_a q$$
$$= (q-1)(1 + 2q + 3q^2 + \cdots + nq^{n-1})\log_a q.$$

But[17]

$$1 + 2q + 3q^2 + \cdots + nq^{n-1} = \frac{nq^n}{q-1} - \frac{q^n - 1}{(q-1)^2},$$

so that

$$S_n = nq^n \log_a q - \frac{q^n - 1}{q-1}\log_a q,$$

or, since $q = \sqrt[n]{b}$, $\log_a q = (1/n)\log_a b$,

$$S_n = b\log_a b - \frac{(b-1)\log_a b}{n(\sqrt[n]{b} - 1)}.$$

Since $\lim_{n\to\infty} n(\sqrt[n]{b} - 1) = \ln b$ (see problem 158c), we have

$$S = \lim_{n\to\infty} S_n = b\log_a b - (b-1)\frac{\log_a b}{\ln b}$$
$$= \frac{b\ln b}{\ln a} - \frac{b-1}{\ln a} \text{ (see footnote 16).}$$

Remark. If we determine the area in part **b** without using the solution to part **a**, as we did in our second solution, then one can use the resulting formula to find the result of part **a**.

160. The solution to this problem is analogous to the second solution of problem 159**b**. Suppose $OA = 1$, $OB = b$. We divide the segment AB into n parts by points $M_1, M_2, \ldots, M_n$ so that $OM_1/1 = OM_2/OM_1 = \cdots = OM_{n-1}/b$ (fig. 107). Then the lengths of the bases of the rectangles shown in the figure are

$$(q-1), q(q-1), q^2(q-1), \ldots, q^{n-1}(q-1),$$

[17] For

$$1 + 2q + 3q^2 + \cdots + nq^{n-1} = (1 + q + q^2 + \cdots + q^{n-1}) + (q + q^2 + \cdots + q^{n-1}) + (q^2 + \cdots + q^{n-1}) + \cdots + (q^{n-2} + q^{n-1}) + q^{n-1}$$
$$= \frac{q^n - 1}{q-1} + \frac{q^n - q}{q-1} + \frac{q^n - q^2}{q-1} + \cdots + \frac{q^n - q^{n-1}}{q-1}$$
$$= \frac{nq^n}{q-1} - \frac{1 + q + \cdots + q^{n-1}}{q-1}$$
$$= \frac{nq^n}{q-1} - \frac{q^n - 1}{(q-1)^2}.$$

Another proof of this identity is given in problem 172**b**.

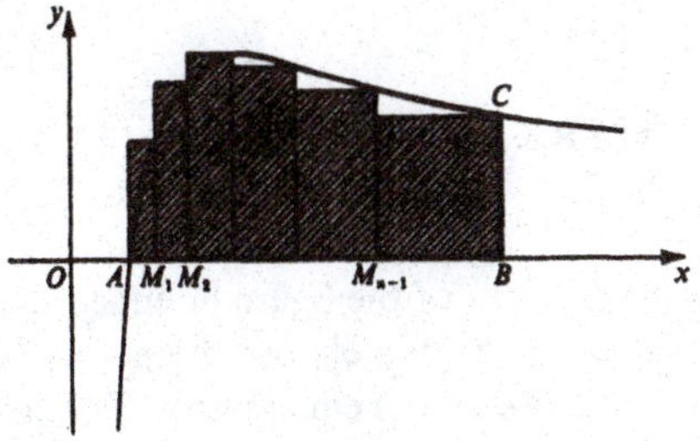

Fig. 107

where $q = \sqrt[n]{b}$, while their heights are

$$\frac{\log_a q}{q}, \quad \frac{2 \log_a q}{q^2}, \quad \frac{3 \log_a q}{q^3}, \ldots, \frac{n \log_a q}{q^n}.$$

(See the solution to 163b.) It follows that the area S of the curvilinear triangle ABC is the limit as $n \to \infty$ of the sum

$$S_n = \frac{(q-1)\log_a q}{q} + \frac{2(q-1)\log_a q}{q} + \cdots + \frac{n(q-1)\log_a q}{q}$$

$$= \frac{q-1}{q}(1 + 2 + \cdots + n)\log_a q = \frac{n(n+1)}{2}\frac{q-1}{q}\log_a q.$$

Since

$$q = \sqrt[n]{b}, \qquad \log_a q = \frac{1}{n}\log_a b,$$

we have

$$S_n = \frac{n+1}{2}\frac{\sqrt[n]{b}-1}{\sqrt[n]{b}}\log_a b = \frac{1}{2}n(\sqrt[n]{b}-1)\frac{n+1}{n}\frac{1}{\sqrt[n]{b}}\log_a b.$$

Now by the result of problem 158c,

$$\lim_{n \to \infty} n(\sqrt[n]{b}-1) = \ln b,$$

and clearly (see footnote, page 140)

$$\lim_{n \to \infty}\frac{n+1}{n} = \lim_{n \to \infty}\left(1 + \frac{1}{n}\right) = 1, \qquad \lim_{n \to \infty}\sqrt[n]{b} = 1.$$

Thus (see footnote, page 159)

$$S = \lim_{n \to \infty} S_n = \tfrac{1}{2}\ln b \log_a b = \tfrac{1}{2}\ln a(\log_a b)^2.$$

In particular, the area of the curvilinear triangle bounded by the x axis, the curve $y = (\ln x)/x$, and the line $x = b$ is equal to $\frac{1}{2}\ln^2 b$, and the area of the curvilinear triangle bounded by the same lines and the curve $y = (\log x)/x$ is equal to

$$\tfrac{1}{2}\ln 10 \log^2 b \approx 1.15 \log^2 b.$$

XII. SOME REMARKABLE LIMITS

161. Let us determine the area of the figure bounded by the curve $y = x^k$, the x axis, and the line $x = b$. We choose the points $M_1, M_2, \ldots, M_{n-1}$ (see p. 26) so as to divide OD into n equal parts. The sum S_n is calculated as in the solution to problem 148: it is

$$S_n = h[h^k + (2h)^k + \cdots + (nh)^k] = h^{k+1}(1^k + 2^k + \cdots + n^k),$$

where $h = b/n$. Thus the required area is

$$S = \lim_{n\to\infty} b^{k+1}\frac{1^k + 2^k + \cdots + n^k}{n^{k+1}}.$$

But from the solution to problem 150**b** we know that $S = b^{k+1}/(k+1)$. It follows that if $k > -1$ (see the note to the solution of 150**b**), then

$$\lim_{n\to\infty}\frac{1^k + 2^k + \cdots + n^k}{n^{k+1}} = \frac{1}{k+1}.$$

Remark. When k is a positive integer, this result can be proved more easily. We give a proof by induction. For $k = 0$ we have

$$\frac{1^0 + 2^0 + \cdots + n^0}{n} = 1 = \frac{1}{0+1},$$

for all n, so that the result certainly holds. Let us write $S_r(n)$ for the sum $1^r + 2^r + \cdots + n^r$. We assume that

$$\lim_{n\to\infty}\frac{S_r(n)}{n^{r+1}} = \frac{1}{r+1} \qquad (0 \leqq r \leqq k-1),$$

and must then prove that

$$\lim_{n\to\infty}\frac{S_k(n)}{n^{k+1}} = \frac{1}{k+1}.$$

Consider the expression $(n+1)^{k+1} - n^{k+1}$. By the binomial theorem we have

$$(n+1)^{k+1} - n^{k+1} = (k+1)n^k + \binom{k+1}{2} n^{k-1} + \cdots + (k+1)n + 1.$$

Similarly, replacing n by $n - 1, n - 2, \ldots, 0$, we have

$$n^{k+1} - (n-1)^{k+1} = (k+1)(n-1)^k + \binom{k+1}{2}(n-1)^{k-1} + \cdots + (k+1)(n-1) + 1$$

$$(n-1)^{k+1} - (n-2)^{k+1} = (k+1)(n-2)^k + \binom{k+1}{2}(n-2)^{k-1} + \cdots + (k+1)(n-2) + 1$$

$$\vdots$$

$$1^{k+1} - (0)^{k+1} = (k+1)\cdot 0^k + \binom{k+1}{2} 0^{k-1} + \cdots + (k+1)\cdot 0 + 1.$$

On adding all these equations, we obtain

$$(n+1)^{k+1} = (k+1)S_k(n) + \binom{k+1}{2} S_{k-1}(n) + \cdots + (k+1)S_1(n) + S_0(n).$$

Now divide through by n^{k+1}, getting

$$\left(\frac{n+1}{n}\right)^{k+1} = (k+1)\frac{S_k(n)}{n^{k+1}} + \binom{k+1}{2}\frac{S_{k-1}(n)}{n^k}\cdot\frac{1}{n} + \cdots + (k+1)\frac{S_1(n)}{n^2}\cdot\frac{1}{n^{k-1}} + \frac{S_0(n)}{n}\cdot\frac{1}{n^k}.$$

Let $n \to \infty$, and use the induction hypothesis, that $\lim_{n\to\infty} S_r(n)/n^{r+1} = 1/(r+1)$ $(0 \leqq r \leqq k-1)$. We obtain

$$1 = \lim_{n\to\infty}\left(\frac{n+1}{n}\right)^{k+1} = (k+1)\lim_{n\to\infty}\frac{S_k(n)}{n^{k+1}} + 0 + 0 + \cdots + 0,$$

or

$$\lim_{n\to\infty}\frac{S_k(n)}{n^{k+1}} = \frac{1}{k+1},$$

as required.

162a. *First solution.* It is sufficient to prove the inequality

$$\log\left[\left(1+\frac{1}{n}\right)^n\right] > \log\left[\left(1+\frac{1}{m}\right)^m\right]$$

or

$$n\log\left(1+\frac{1}{n}\right) > m\log\left(1+\frac{1}{m}\right)$$

whenever $n > m$. We take logarithms to base e (see problems 151 to 154 above). Then $\ln(1 + 1/n)$ is the area of the curvilinear trapezoid $ABDC$ bounded by the hyperbola $y = 1/x$, the x axis, and the lines $x = 1$ and $x = 1 + 1/n$; $\ln(1 + 1/m)$ is defined similarly (fig. 108).

Let $ABNK$ be the rectangle with base AB and area equal to that of the curvilinear trapezoid $ABCD$; then

$$S(ABNK) = AB \cdot AK = \frac{1}{n} \cdot AK,$$

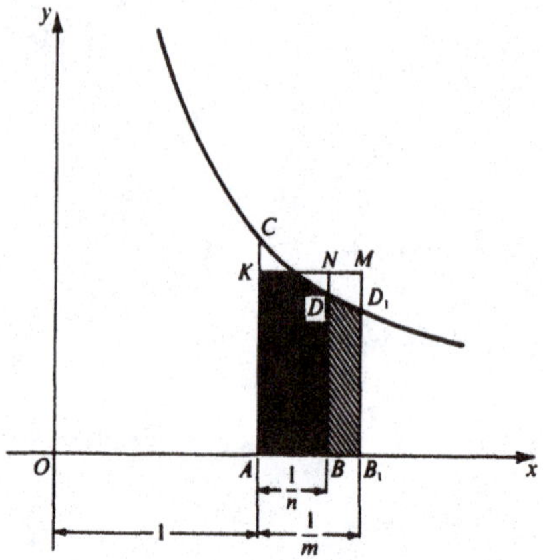

Fig. 108

and therefore

$$\frac{1}{n} \cdot AK = S(ABDC) = \ln\left(1 + \frac{1}{n}\right), \qquad AK = n \ln\left(1 + \frac{1}{n}\right).$$

Next, using the notation of fig. 108, $S(ABNK) = S(ABDC)$ and $S(BB_1MN) > S(BB_1D_1D)$, so that $S(AB_1MK) > S(AB_1D_1C)$. But

$$S(AB_1MK) = AB_1 \cdot AK = \frac{1}{m} AK$$

$$= \frac{1}{m} n \ln\left(1 + \frac{1}{n}\right), \qquad S(AB_1D_1C) = \ln\left(1 + \frac{1}{m}\right),$$

so that

$$\frac{1}{m} n \ln\left(1 + \frac{1}{n}\right) > \ln\left(1 + \frac{1}{m}\right),$$

$$n \ln\left(1 + \frac{1}{n}\right) > m \ln\left(1 + \frac{1}{m}\right),$$

as required.

Second solution. We use the *inequality between the arithmetic and geometric means*,[18] which states that if $a_1, a_2, \ldots, a_n$ are all $\geqq 0$, then

$$\sqrt[n]{a_1 a_2 \cdots a_n} \leqq \frac{a_1 + a_2 + \cdots + a_n}{n}, \tag{1}$$

with strict inequality unless $a_1 = a_2 = \cdots = a_n$. Suppose $n > 1$, and put

$$a_1 = a_2 = \cdots = a_{n-1} = 1 + \frac{1}{n-1}, \qquad a_n = 1.$$

Since the a_i are not all equal, the strict inequality holds, and we get

$$\sqrt[n]{\left(1 + \frac{1}{n-1}\right)^{n-1}} < \frac{(n-1)\left(1 + \dfrac{1}{n-1}\right) + 1}{n}$$
$$= 1 + \frac{1}{n}.$$

Raising both sides to the nth power, it follows that

$$\left(1 + \frac{1}{n-1}\right)^{n-1} < \left(1 + \frac{1}{n}\right)^n,$$

which is what we wanted to prove.

Third solution. Expand $\left(1 + \frac{1}{n}\right)^n$ by the binomial theorem; we get

$$\left(1 + \frac{1}{n}\right)^n = 1 + 1 + \frac{n(n-1)}{2}\frac{1}{n^2} + \frac{n(n-1)(n-2)}{6}\frac{1}{n^3} + \cdots + \frac{1}{n^n}.$$

The number of terms is $n + 1$, and thus increases with n. Moreover, the individual terms increase with n, since

$$\frac{n(n-1)(n-2)\cdots(n-k+1)}{k!}\frac{1}{n^k}$$
$$= \frac{1}{k!}\left(1 - \frac{1}{n}\right)\left(1 - \frac{2}{n}\right)\cdots\left(1 - \frac{k-1}{n}\right),$$

and each factor $1 - j/n$ is an increasing function of n. Hence $\left(1 + \frac{1}{n}\right)^n$ increases with n.

[18] The left-hand side of (1) is the geometric mean of the numbers $a_1, a_2, \ldots a_n$, while the right side is their arithmetic mean. For a proof of the inequality, see for example, Ref. [3].

b. *First solution.* It is sufficient to show that for any positive integer n

$$\ln\left[\left(1+\frac{1}{n}\right)^{n+1}\right] > \ln\left[\left(1+\frac{1}{n+1}\right)^{n+2}\right].$$

Form the difference

$$\ln\left[\left(1+\frac{1}{n}\right)^{n+1}\right] - \ln\left[\left(1+\frac{1}{n+1}\right)^{n+2}\right]$$
$$= (n+1)\ln\left(1+\frac{1}{n}\right) - (n+2)\ln\left(1+\frac{1}{n+1}\right)$$
$$= (n+1)\left[\ln\left(1+\frac{1}{n}\right) - \ln\left(1+\frac{1}{n+1}\right)\right] - \ln\left(1+\frac{1}{n+1}\right).$$

But $\ln(1 + 1/n) - \ln(1 + 1/(n+1))$ is the area of the curvilinear trapezoid BB_1D_1D bounded by the hyperbola $y = 1/x$, the x axis, and the lines $x = 1 + 1/n$, $x = 1 + 1/(n+1)$, and $\ln(1 + 1/(n+1))$ is the area of

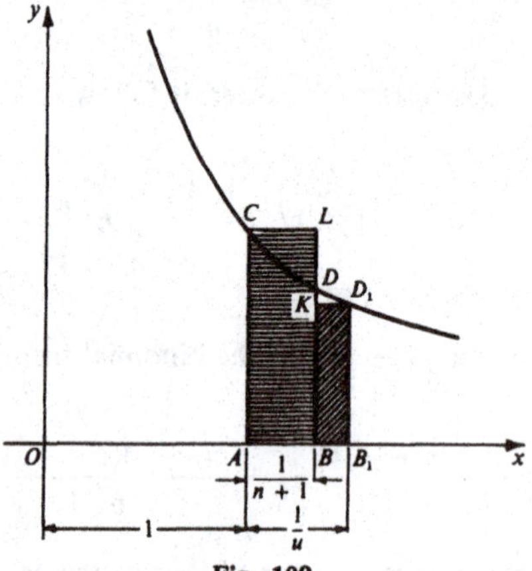

Fig. 109

the curvilinear trapezoid bounded by the same curve, the x axis, and the lines $x = 1$ and $x = 1 + 1/(n+1)$ (fig. 109). The area of the curvilinear trapezoid BB_1D_1D is greater than the area of the rectangle BB_1D_1K, whose area is

$$BB_1 \cdot B_1D_1 = \left(\frac{1}{n} - \frac{1}{n+1}\right)\frac{1}{1+1/n} = \frac{1}{n(n+1)}\,\frac{n}{n+1} = \left(\frac{1}{n+1}\right)^2.$$

Thus

$$(n+1)\left[\ln\left(1+\frac{1}{n}\right) - \ln\left(1+\frac{1}{n+1}\right)\right] > (n+1)\cdot\frac{1}{(n+1)^2}$$
$$= \frac{1}{n+1}.$$

The area of the curvilinear trapezoid $ABCD$ is less than the area of the recta ıgle $ABLC$, which is equal to

$$AB \cdot AC = \frac{1}{n+1} \cdot 1 = \frac{1}{n+1}.$$

Thus

$$\ln\left(1 + \frac{1}{n+1}\right) < \frac{1}{n+1}.$$

On comparing the two inequalities, we find that

$$\ln\left[\left(1 + \frac{1}{n}\right)^{n+1}\right] - \ln\left[\left(1 + \frac{1}{n+1}\right)^{n+2}\right]$$
$$= (n-1)\left[\ln\left(1 + \frac{1}{n}\right) - \ln\left(1 + \frac{1}{n+1}\right)\right] - \ln\left(1 + \frac{1}{n+1}\right) > 0,$$

as required.

Second solution. As in the second solution to part **a**, we can use the inequality between the arithmetic and geometric means. Take

$$a_1 = a_2 = \cdots = a_n = 1 - \frac{1}{n}, \ a_{n+1} = 1, \text{ where } n > 1.$$

Then we have the strict inequality, so

$$\sqrt[n+1]{\left(1 - \frac{1}{n}\right)^n} < \frac{n\left(1 - \frac{1}{n}\right) + 1}{n+1} = \frac{n}{n+1}.$$

Hence

$$\left(\frac{n-1}{n}\right)^n < \left(\frac{n}{n+1}\right)^{n+1}.$$

Inverting both sides, we reverse the inequality; thus

$$\left(\frac{n}{n-1}\right)^n > \left(\frac{n+1}{n}\right)^{n+1}.$$

In other words,

$$\left(1 + \frac{1}{n-1}\right)^n > \left(1 + \frac{1}{n}\right)^{n+1},$$

which is what we wanted to prove.

Remark. In later problems it will be important to know that

$$\lim_{n\to\infty}\left(1 + \frac{1}{n}\right)^n = e = \lim_{n\to\infty}\left(1 + \frac{1}{n}\right)^{n+1},$$

where e is the number defined geometrically in problem 153. This follows at once from either problem 158**a** or problem 163 (with $z = 1$).

163. First suppose that z is positive. Consider the curvilinear trapezoid $ABCD$ bounded by the hyperbola $y = 1/x$, the x axis, and the lines $x = 1$ and $x = 1 + z/n$ (fig. 110a). The area of this trapezoid lies between the areas of the rectangles $ABED$ and $AFCD$, that is, between $z/n \cdot 1$ and $z/n \cdot 1/(1 + z/n) = z/(n + z)$ (see the end of the solution of 159). It follows that

$$\frac{z}{n} > \ln\left(1 + \frac{z}{n}\right) > \frac{z}{n+z}$$

or

$$z > \ln\left[\left(1 + \frac{z}{n}\right)^n\right] > \frac{z}{1 + z/n}. \tag{1}$$

Passing to the limit[19] as $n \to \infty$, we see that

$$\ln\left[\lim_{n\to\infty}\left(1 + \frac{z}{n}\right)^n\right] = z, \tag{2}$$

from which it follows that

$$\lim_{n\to\infty}\left(1 + \frac{z}{n}\right)^n = e^z.$$

The case where z is negative is handled similarly (fig. 110b). Here $S(ABED) = -z/n$, $S(AFCD) = \left(-\frac{z}{n}\right)\frac{1}{1 + \frac{z}{n}} = \frac{-z}{n + z}$, $S(ABCD) = -\ln(1 + (z/n))$. (Remember that z is negative; see p. 31.) It follows from fig. 110b that

$$-\frac{z}{n} < -\ln\left(1 + \frac{z}{n}\right) < \frac{-z}{n+z},$$

or

$$\frac{z}{n} > \ln\left(1 + \frac{z}{n}\right) > \frac{z}{n+z},$$

or finally,

$$z > \ln\left[\left(1 + \frac{z}{n}\right)^n\right] > \frac{z}{1 + z/n}.$$

[19] The details are as follows. From (1) we see that $(1 + z/n)^n < e^z$, so the sequence $(1 + z/n)^n$ is bounded from above. As in problem 162, it can be shown that the sequence $(1 + z/n)^n$ is increasing. Hence $\lim (1 + z/n)^n$ exists. Since $\ln x$ is continuous, we have $\lim\limits_{n\to\infty} \ln (1 + z/n)^n = \ln \lim\limits_{n\to\infty} (1 + z/n)^n$. Since the first and last members of (1) tend to z as $n \to \infty$, so must its middle member, i.e., $\lim\limits_{n\to\infty} \ln (1 + z/n)^n = z$. Equation (2) follows at once from these facts.

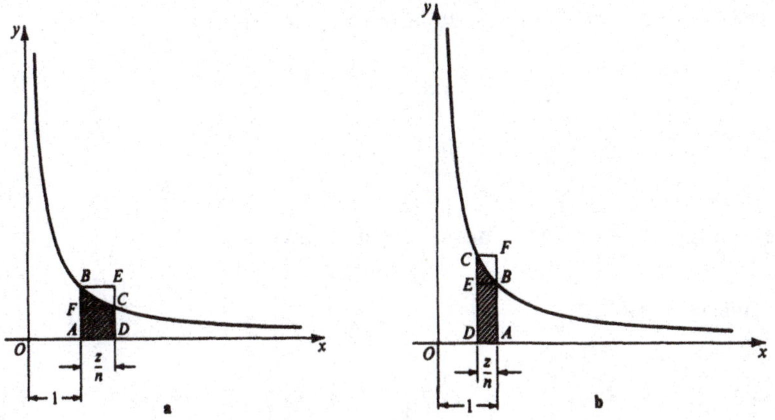

Fig. 110

Letting $n \to \infty$, we find once again

$$\ln\left[\lim_{n\to\infty}\left(1+\frac{z}{n}\right)^n\right] = z,$$

and so

$$\lim_{n\to\infty}\left(1+\frac{z}{n}\right)^n = e^z.$$

164. Expand $(1 + z/n)^n$ by the binomial theorem:

$$\left(1+\frac{z}{n}\right)^n = 1 + \frac{z}{n}\cdot\frac{n}{1} + \frac{z^2}{n^2}\cdot\frac{n(n-1)}{2!}$$

$$+\cdots+\frac{z^n}{n^n}\frac{n(n-1)(n-2)\cdots(n-\overline{n-1})}{n!}$$

$$= 1 + z + \frac{z^2}{2!}\left(1-\frac{1}{n}\right) + \frac{z^3}{3!}\left(1-\frac{1}{n}\right)\left(1-\frac{z}{n}\right)$$

$$+\cdots+\frac{z^n}{n!}\left(1-\frac{1}{n}\right)\left(1-\frac{2}{n}\right)\cdots\left(1-\frac{n-1}{n}\right).$$

We can write

$$\left(1+\frac{z}{n}\right)^n = 1 + u_1 + u_2 + \cdots + u_n,$$

where

$$u_k = \frac{z^k}{k!}\left(1-\frac{1}{n}\right)\left(1-\frac{2}{n}\right)\cdots\left(1-\frac{k-1}{n}\right) \qquad (k = 1, 2, \ldots, n).$$

The terms u_k of this sum clearly satisfy the inequalities

$$|u_k| \leqq \frac{|z|^k}{k!},$$

$$\left|\frac{u_{k+1}}{u_k}\right| = \frac{|z|\,[1-(k/n)]}{k+1} \leqq \frac{|z|}{k+1}.$$

From the second of these inequalities it follows that

$$|u_{k+1}| \leqq |u_k| \cdot \frac{|z|}{k+1}; \qquad |u_{k+2}| < |u_{k+1}| \cdot \frac{|z|}{k+2} \leqq |u_k| \cdot \frac{|z|^2}{(k+1)^2};$$

$$|u_{k+3}| \leqq |u_{k+2}| \frac{|z|}{k+3} \leqq |u_k| \cdot \frac{|z|^3}{(k+1)^3}, \ldots .$$

Now choose k so that $k + 1 > |z|$. This is permissible, since we shall later let $n \to \infty$, and our sums can therefore be assumed to have sufficiently many terms. For such k

$$\begin{aligned} |u_{k+1} + u_{k+2} + \cdots + u_n| &\leqq |u_{k+1}| + |u_{k+2}| + \cdots + |u_n| \\ &\leqq |u_k| \left[\frac{|z|}{k+1} + \left(\frac{|z|}{k+1}\right)^2 + \cdots + \left(\frac{|z|}{k+1}\right)^{n-k}\right] \\ &\leqq |u_k| \frac{|z|/(k+1)}{1 - |z|/(k+1)} = |u_k| \frac{|z|}{k+1-|z|} \\ &\leqq \frac{|z|^{k+1}}{k!\,(k+1-|z|)}. \end{aligned}$$

But

$$\left(1 + \frac{z}{n}\right)^n - (1 + u_1 + u_2 + \cdots + u_k) = u_{k+1} + u_{k+2} + \cdots + u_n,$$

so that

$$\left|\left(1 + \frac{z}{n}\right)^n - (1 + u_1 + u_2 + \cdots + u_k)\right| \leqq \frac{|z|^{k+1}}{k!\,(k+1-|z|)}.$$

It follows that

$$\lim_{n\to\infty} \left|\left(1 + \frac{z}{n}\right)^n - (1 + u_1 + u_2 + \cdots + u_k)\right| \leqq \frac{|z|^{k+1}}{k!\,(k+1-|z|)}.$$

Now $\lim_{n\to\infty} u_k = z^k/k!$ and $\lim_{n\to\infty} (1 + z/n)^n = e^z$ (see problem 163). Hence

$$\left|e^z - \left(1 + z + \frac{z^2}{2!} + \cdots + \frac{z^k}{k!}\right)\right| \leqq \frac{|z|^{k+1}}{k!\,(k+1-|z|)}.$$

But clearly

$$\lim_{k\to\infty} \frac{|z|^{k+1}}{k!\,(k+1-|z|)} = 0.$$

Thus

$$\lim_{k\to\infty} \left[e^z - \left(1 + z + \frac{z^2}{2!} + \frac{z^3}{3!} + \cdots + \frac{z^k}{k!}\right)\right] = 0,$$

and this is exactly what is meant by the equation

$$e^z = 1 + \frac{z}{1!} + \frac{z^2}{2!} + \frac{z^3}{3!} + \cdots + \frac{z^n}{n!} + \cdots.$$

165. Consider the sum of the areas of the $n - 2$ trapezoids and the triangle inscribed in the curve $y = \ln x$ in fig. 111a. Since the bases of the kth trapezoid are $\ln k$ and $\ln (k + 1)$, this sum is equal to

$$\frac{\ln 2}{2} + \frac{\ln 3 + \ln 2}{2} + \frac{\ln 4 + \ln 3}{2} + \cdots + \frac{\ln n + \ln (n-1)}{2}$$

$$= \ln 2 + \ln 3 + \ln 4 + \cdots + \ln (n-1) + \frac{1}{2} \ln n.$$

The area S_n bounded by the curve $y = \ln x$, the x axis, and the line $x = n$ is clearly greater than the area of the inscribed figure. That is,

$$S_n > \ln 2 + \ln 3 + \cdots + \ln (n-1) + \tfrac{1}{2} \ln n.$$

Moreover,

$$K_n = S_n - [\ln 2 + \ln 3 + \ln 4 + \cdots + \ln (n-1) + \tfrac{1}{2} \ln n]$$

is an increasing function of n, since this difference is equal to the sum of the areas of the $n - 1$ segments cut from the curve by the upper boundaries of the trapezoids, and as n increases new segments are added.

Let us now construct $n - 2$ trapezoids, bounded by lines $x = k - \frac{1}{2}$ and $x = k + \frac{1}{2}$, the tangent to the curve at the point $x = k$ and the x axis ($k = 2, 3, 4, \ldots, n - 1$). Add a further trapezoid bounded by the lines $x = 1$ and $x = \frac{3}{2}$, the tangent to the curve at $x = \frac{5}{4}$ and the x axis, and a

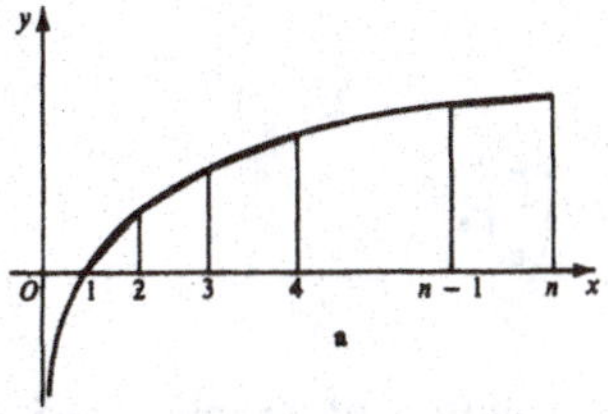

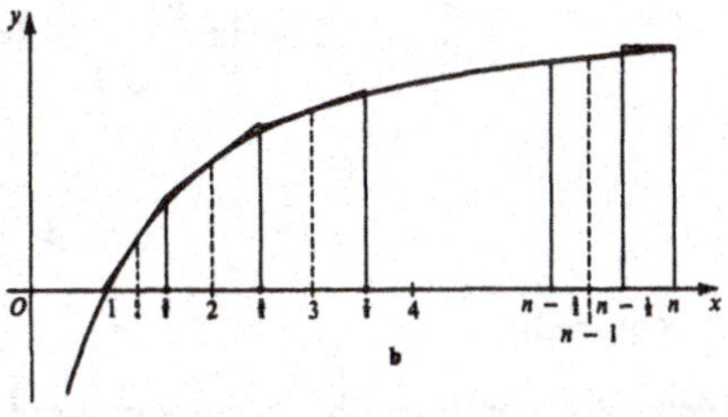

Fig. 111

rectangle of height $\ln n$ with sides on $x = n - \frac{1}{2}$ and $x = n$ (fig. 111b). Since the midline of the kth trapezoid is of length $\ln(k+1)$, and the midline of the small added trapezoid of length $\ln \frac{5}{4}$, the area of the entire figure is

$$[\ln 2 + \ln 3 + \cdots + \ln(n-1)] + \tfrac{1}{2}\ln n + \tfrac{1}{2}\ln \tfrac{5}{4}$$
$$= [\ln 2 + \ln 3 + \cdots + \ln(n-1) + \tfrac{1}{2}\ln n] + \tfrac{1}{2}\ln \tfrac{5}{4}.$$

But this area is greater than that of S_n, so that

$$[\ln 2 + \ln 3 + \cdots + \ln(n-1) + \tfrac{1}{2}\ln n] + \tfrac{1}{2}\ln \tfrac{5}{4} > S_n.$$

It follows that for any n,

$$K_n < \tfrac{1}{2}\ln \tfrac{5}{4}.$$

In the solution of problem 159**b** we showed that

$$S_n = n \ln n - n + 1.$$

On the other hand,

$$\ln 2 + \ln 3 + \cdots + \ln(n-1) + \ln n = \ln(2 \cdot 3 \cdots n) = \ln n!$$

Using these relations, we can rewrite the expression for K_n in the following form:

$$K_n = (n \ln n - n + 1) - \left(\ln n! - \frac{1}{2}\ln n\right)$$
$$= \ln n^n - \ln e^n + 1 - \ln n! + \ln\sqrt{n} = \ln \frac{n^n\sqrt{n}}{n!\, e^n} + 1.$$

Since $0 < K_n < \frac{1}{2}\ln \frac{5}{4}$ for all n,

$$0 < \ln \frac{n^n\sqrt{n}}{n!\, e^n} + 1 < \frac{1}{2}\ln\frac{5}{4},$$

that is,

$$-1 < \ln \frac{n^n\sqrt{n}}{n!\, e^n} < -1 + \frac{1}{2}\ln\frac{5}{4},$$

or

$$1 > \ln \frac{n!\, e^n}{n^n\sqrt{n}} > \ln\sqrt{\frac{4}{5}}\, e$$

or

$$e\sqrt{n}\left(\frac{n}{e}\right)^n > n! > \sqrt{\frac{4}{5}}\, e\sqrt{n}\left(\frac{n}{e}\right)^n,$$

as required.

166a. We continue the argument used in the solution of problem 165. Since K_n increases with increasing n, yet always remains less than $\frac{1}{2}\ln \frac{5}{4}$, it must have a limit, also not exceeding $\frac{1}{2}\ln\frac{5}{4}$, but necessarily positive, since K_2 is positive, and all subsequent K's are larger. Thus

$$\lim_{n\to\infty} K_n = K, \qquad \text{where } 0 < K \leqq \frac{1}{2}\ln\frac{5}{4}.$$

But in the solution to 165 we showed that

$$K_n = \ln\frac{n^n\sqrt{n}}{n!\,e^n} + 1.$$

Since the limit $K = \lim_{n\to\infty} K_n$ exists, so does the limit

$$C = \lim_{n\to\infty}\frac{n!\,e^n}{n^n\sqrt{n}},$$

and moreover,

$$e > C \geqq \sqrt{\tfrac{5}{4}}\,e,$$

as required.

b. In order to determine the numerical value of the limit C as $n\to\infty$ of the quantity

$$\frac{n!}{\sqrt{n}\left(\frac{n}{e}\right)^n} = C_n,$$

we use Wallis's formula (problem 147):

$$\frac{\pi}{2} = \lim_{n\to\infty}\frac{2}{1}\cdot\frac{2}{3}\cdot\frac{4}{3}\cdot\frac{4}{5}\cdots\frac{2n}{2n-1}\cdot\frac{2n}{2n+1},$$

which may be written in the form

$$\frac{\pi}{2} = \lim_{n\to\infty}\frac{(2\cdot4\cdot6\cdots2n)^4}{(1\cdot2\cdot3\cdot4\cdot5\cdots2n)^2(2n+1)} = \lim_{n\to\infty}\frac{[2^n(n!)]^4}{[(2n)!]^2(2n+1)},$$

or

$$\sqrt{\pi} = \lim_{n\to\infty}\frac{2^{2n}(n!)^2}{(2n)!\cdot\sqrt{\dfrac{2n+1}{2}}}.$$

In this last equality we substitute

$$n! = C_n\sqrt{n}\left(\frac{n}{e}\right)^n, \qquad (2n)! = C_{2n}\sqrt{2n}\left(\frac{2n}{e}\right)^{2n}$$

and obtain

$$\sqrt{\pi} = \lim_{n\to\infty}\frac{2^{2n}C_n^{\,2}n\left(\dfrac{n}{e}\right)^{2n}}{C_{2n}\sqrt{2n}\left(\dfrac{2n}{e}\right)^{2n}\sqrt{\dfrac{2n+1}{2}}} = \lim_{n\to\infty}\frac{C_n^{\,2}}{C_{2n}\sqrt{\dfrac{2n+1}{n}}}.$$

But

$$\lim_{n\to\infty} C_n = \lim_{n\to\infty} C_{2n} = C,$$

$$\lim_{n\to\infty} \sqrt{\frac{2n+1}{n}} = \lim_{n\to\infty} \sqrt{2 + \frac{1}{n}} = \sqrt{2},$$

so that

$$\sqrt{\pi} = \frac{C^2}{C\sqrt{2}},$$

whence

$$C = \sqrt{2\pi},$$

as required.

167a. ln n is equal to the area of the curvilinear trapezoid $ABCD$ bounded by the hyperbola $y = 1/x$, the x axis, and the lines $x = 1$ and $x = n$.

Construct n rectangles, all of base 1 and heights $1, \frac{1}{2}, \frac{1}{3}, \ldots, 1/(n-1)$, as in fig. 112a. The area of the curvilinear trapezoid $ABCD$ is less than

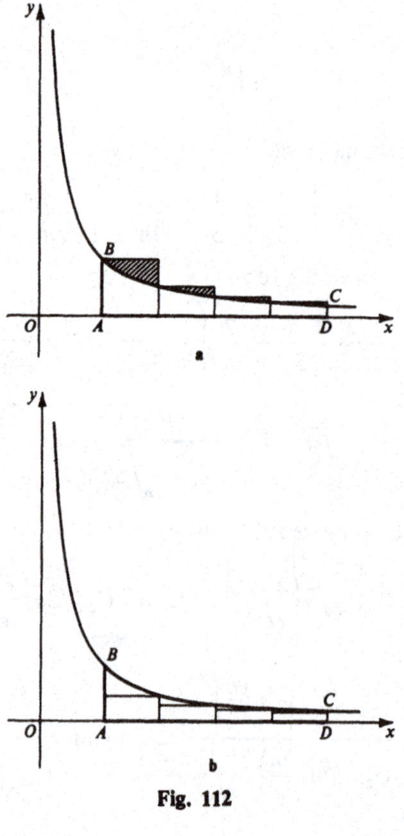

Fig. 112

the total area of these rectangles. Hence

$$1 + \frac{1}{2} + \frac{1}{3} + \cdots + \frac{1}{n-1} > \ln n,$$

that is,

$$\gamma_n = 1 + \frac{1}{2} + \frac{1}{3} + \cdots + \frac{1}{n-1} - \ln n > 0.$$

We now inscribe in the same curvilinear trapezoid n rectangles with unit bases and heights $\frac{1}{2}, \frac{1}{3}, \frac{1}{4}, \ldots, 1/n$ (fig. 112b). Since the total area of these rectangles is less than that of $ABCD$,

$$\ln n > \frac{1}{2} + \frac{1}{3} + \frac{1}{4} + \cdots + \frac{1}{n}.$$

Therefore

$$\gamma_n = 1 + \frac{1}{2} + \frac{1}{3} + \cdots + \frac{1}{n-1} - \ln n < 1 - \frac{1}{n} < 1.$$

Thus for any n, $0 < \gamma_n < 1$.

b. As n increases, so does γ_n, for it is equal to the sum of the shaded areas in fig. 112a. Since with increasing n,

$$\gamma_n = 1 + \frac{1}{2} + \frac{1}{3} + \cdots + \frac{1}{n-1} - \ln n$$

increases, yet always remains less than 1, it must have a limit γ, satisfying $0 < \gamma \leqq 1$.

168a. The sum

$$\frac{\log 1}{1} + \frac{\log 2}{2} + \frac{\log 3}{3} + \cdots + \frac{\log (n-1)}{n-1}$$

is equal to the area of the shaded figure shown in fig. 113. For large n this area does not differ greatly from the area of the curvilinear triangle ABC, which by problem 160 is

$$\frac{\ln 10}{2} \log^2 n.$$

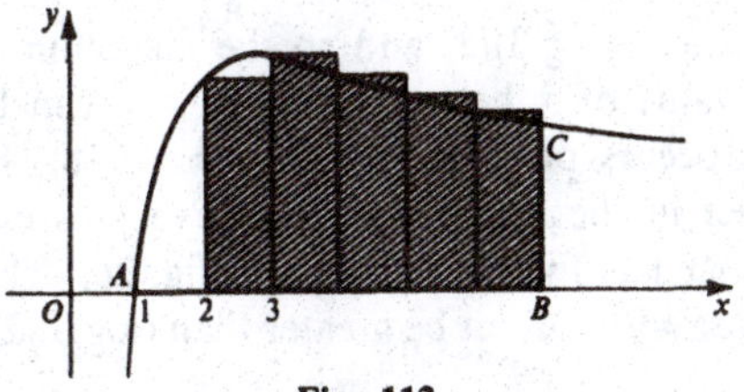

Fig. 113

It is therefore natural to suppose that the required number C is (ln 10)/2. We consider, then, the difference

$$\frac{\log 1}{1} + \frac{\log 2}{2} + \cdots + \frac{\log n}{n} - C \log^2 n,$$

where $C = (\ln 10)/2$, but first let us examine in greater detail the form of the curve $y = (\log x)/x$.

Clearly, as $x \to 0$ the quantity $(\log x)/x$ tends to $-\infty$, for $x = 1$ it is zero, and as $x \to \infty$ it tends to zero. This last follows from the fact that the ratio of the number of digits in a number n (written out in the usual way to base 10) and n itself tends to zero as n increases, since the number of digits in n does not differ from $\log n$ by more than 1. Now we use the fact that

$$\frac{\sqrt[n+1]{n+1}}{\sqrt[n]{n}} = \sqrt[n(n+1)]{\frac{(n+1)^n}{n^{n+1}}} = \sqrt[n(n+1)]{\left(1+\frac{1}{n}\right)^n \frac{1}{n}}$$
$$< \sqrt[n(n+1)]{\frac{e}{n}}$$

(see problem 162). It follows for $n \geqq 3$ that $\sqrt[n+1]{n+1} < \sqrt[n]{n}$. But $\sqrt{2} = 1.41$, and $\sqrt[3]{3} = 1.44\ldots$, so that $\sqrt{2} < \sqrt[3]{3}$ and $\sqrt[3]{3} > \sqrt[4]{4} > \sqrt[5]{5} > \sqrt[6]{6} > \ldots$. On taking logarithms, we find that

$$\frac{\log 2}{2} < \frac{\log 3}{3} \quad \text{and} \quad \frac{\log 3}{3} > \frac{\log 4}{4} > \frac{\log 5}{5} > \frac{\log 6}{6} > \cdots.$$

We see therefore that as x increases, the quantity $(\log x)/x$ at first increases, then somewhere between $x = 2$ and $x = 3$ assumes a maximum value, and then decreases. To locate the maximum more closely, we must do a little more calculating.

Let us compare (log 2.5)/2.5 with (log 3)/3. We have

$$\frac{\log 2.5}{2.5} = \frac{\log 10/4}{10/4} = \frac{4(1 - 2\log 2)}{10} = \frac{4(1 - 0.6026\cdots)}{10} = 0.15917\cdots$$

and

$$\frac{\log 3}{3} = \frac{0.47712\cdots}{3} = 0.15904\cdots.$$

Hence $(\log 2.5)/2.5 > (\log 3)/3$, and so the maximum value of $(\log x)/x$ occurs at some value of x between 2 and 3. It can be shown that the maximum value occurs precisely at $x = e \approx 2.718$. However, we shall not need this fact in the argument. We have thus established that the curve $y = (\log x)/x$ has the form shown in fig. 26. The maximum value of the function $(\log x)/x$ cannot be greater than $(\log 3)/2$, for $\log x < \log 3$ for $2 < x < 3$.

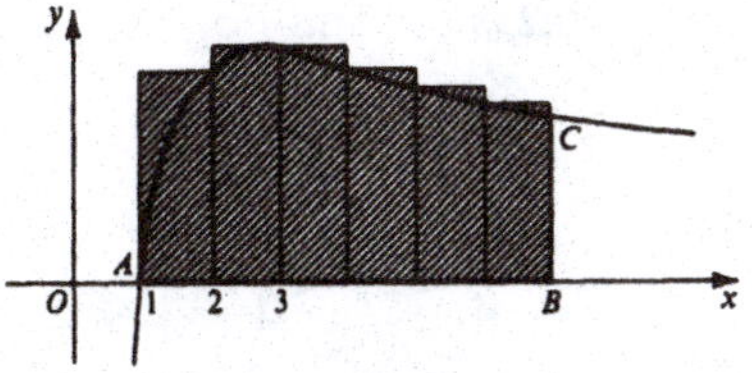

Fig. 114

The area of the curvilinear figure ABC is less than that of the shaded polygon in fig. 114. The area of this shaded figure is less than

$$\frac{\log 2}{2} + \frac{\log 3}{2} + \frac{\log 3}{3} + \frac{\log 4}{4} + \cdots + \frac{\log (n-1)}{n-1}.$$

[For the area of the first rectangle is (log 2)/2, that of the second, less than (log 3)/2, while the areas of the third, fourth, . . . rectangles are (log 3)/3, (log 4)/4, ...] Since (log 1)/1 = 0, we have

$$\frac{\log 1}{1} + \frac{\log 2}{2} + \frac{\log 3}{3} + \frac{\log (n-1)}{n-1} + \frac{\log 3}{2} > C \log^2 n,$$

that is,

$$\delta_n = \frac{\log 1}{1} + \frac{\log 2}{2} + \cdots + \frac{\log (n-1)}{n-1} - C \log^2 n > -\frac{\log 3}{2}.$$

On the other hand, the area of the curvilinear triangle ABC is greater than the area of the curvilinear trapezoid $MDEN$; it is therefore still greater than the area of the shaded figure in fig. 115. The area of this figure is

$$\frac{\log 2}{2} + \frac{\log 4}{4} + \frac{\log 5}{5} + \cdots + \frac{\log (n-1)}{n-1}.$$

Thus

$$\frac{\log 2}{2} + \frac{\log 3}{3} + \frac{\log 4}{4} + \cdots + \frac{\log (n-1)}{n-1} - \frac{\log 3}{3} < C \log^2 n,$$

where $C = (\ln 10)/2$, that is,

$$\delta_n = \frac{\log 1}{1} + \frac{\log 2}{2} + \frac{\log 3}{3} + \cdots + \frac{\log (n-1)}{n-1} - C \log^2 n < \frac{\log 3}{3}.$$

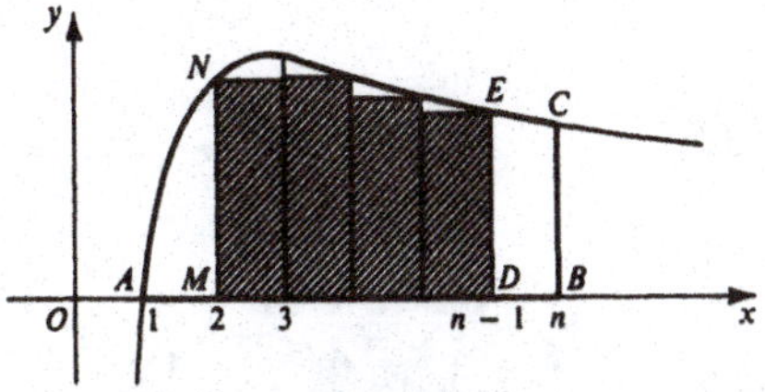

Fig. 115

Since $(-\log 3)/2 = -0.238\ldots$ and $(\log 3)/3 = 0.159\ldots$, we find that

$$-0.239 < \delta_n < 0.16$$

and so certainly

$$-\frac{1}{4} < \delta_n < \frac{1}{4}.$$

b. The difference

$$\varepsilon_n = \frac{\log 2}{2} + \frac{\log 4}{4} + \frac{\log 5}{5} + \cdots + \frac{\log (n-1)}{n-1} - (C \log^2 n - C \log^2 2),$$

where $C = (\ln 10)/2$ and $n \geqq 3$ is equal to the area of that part of the curvilinear trapezoid $MBCN$ which is not shaded in fig. 115. Hence this difference increases with n. By the result of part **a**,

$$\varepsilon_n = \delta_n - \frac{\log 3}{3} + C \log^2 2 < \frac{1}{4} - \frac{\log 3}{3} + C \log^2 2$$

for any value of n. Hence as $n \to \infty$, ε_n tends to a limit ε. Therefore the sequence $\{\delta_n\}$, where $\delta_n = \varepsilon_n + (\log 3)/3 - C \log^2 2$ also tends to a limit $\delta = \varepsilon + (\log 3)/3 - C \log^2 2$.

169. The solution to this problem is analogous to that of 167**a**. Let $ABCD$ be the curvilinear trapezoid bounded by the curve $y = 1/x^s$, the x axis, and the lines $x = 1$ and $x = n$. Consider the polygon consisting of $n - 1$ rectangles, with base 1 inscribed in the trapezoid, and the polygon consisting of $n - 1$ rectangles of base 1 circumscribed about it (see figs. 122a and b). The areas of these polygons are

$$1 + \frac{1}{2^s} + \frac{1}{3^s} + \cdots + \frac{1}{(n-1)^s} \quad \text{and} \quad \frac{1}{2^s} + \frac{1}{3^s} + \cdots + \frac{1}{(n-1)^s} + \frac{1}{n^s},$$

and the area of the curvilinear trapezoid $ABCD$ is $(1 - 1/n^{s-1})/(s - 1)$. (See the solution to 150**a**.) Therefore,

$$\frac{1}{2^s} + \frac{1}{3^s} + \cdots + \frac{1}{(n-1)^s} + \frac{1}{n^s} < \frac{1 - (1/n^{s-1})}{s-1}$$

$$< 1 + \frac{1}{2^s} + \frac{1}{3^s} + \cdots + \frac{1}{(n-1)^s}$$

or

$$\frac{1 - (1/n^{s-1})}{s-1} < 1 + \frac{1}{2^s} + \frac{1}{3^s} + \cdots + \frac{1}{(n-1)^s} < \frac{1 - (1/n^{s-1})}{s-1} + 1 - \frac{1}{n^s},$$

or, finally,

$$\frac{1}{s-1} - \frac{1}{s-1}\frac{1}{n^{s-1}} < 1 + \frac{1}{2^s} + \frac{1}{3^s} + \cdots + \frac{1}{(n-1)^s} < \frac{s}{s-1} - \frac{s}{s-1}\frac{1}{n^{s-1}}.$$

Since $1/n^s \to 0$ as $n \to \infty$, the assertion of the problem follows from this inequality. That the sum $1 + \frac{1}{2^s} + \frac{1}{3^s} + \cdots + \frac{1}{n^s}$ does tend to a definite limit as $n \to \infty$ follows from the fact that it increases with increasing n, yet remains bounded above by $s/(s-1)$.

XIII. THE THEORY OF PRIMES

170. Consider the expression

$$\binom{2n}{n} = \frac{(2n)!}{(n!)^2}.$$

We shall estimate its value by two different methods.

First, by the binomial theorem,

$$1 + \binom{2n}{1} + \binom{2n}{2} + \cdots + \binom{2n}{n} + \cdots + \binom{2n}{2n-1} + 1 = (1+1)^{2n} = 2^{2n},$$

and therefore

$$\binom{2n}{n} < 2^{2n}.$$

Next,

$$\binom{2n}{n} = \frac{2n(2n-1)(2n-2)\ldots(n+1)}{1\cdot 2\cdot 3\ldots n} = \frac{2n}{n}\cdot\frac{2n-1}{n-1}\cdot\frac{2n-2}{n-2}\cdots\frac{n+1}{1}$$
$$\geqq \underbrace{2\cdot 2\cdot 2\cdots 2}_{n\text{ times}} = 2^n$$

so that

$$2^n \leqq \binom{2n}{n} < 2^{2n}. \tag{1}$$

Second, let us consider the prime factorization of $\binom{2n}{n}$. The prime p appears in the factorization of the number $k!$ to the power

$$\left[\frac{k}{p}\right] + \left[\frac{k}{p^2}\right] + \left[\frac{k}{p^3}\right] + \cdots,$$

where brackets indicate integral parts. To see this, observe that among the k factors of $k! = 1\cdot 2\cdot 3\cdots k$, there are $[k/p]$ divisible by p; of these, $[k/p^2]$ are divisible by p^2 (and therefore have to be counted twice); $[k/p^3]$

are divisible by p^3 (and therefore have to be counted three times), and so on. Thus only the primes $p_1, p_2, \ldots, p_r$, where $p_r \leqq 2n < p_{r+1}$, appear in the prime decomposition of $\binom{2n}{n} = (2n)!/(n!)^2$, and each p_i appears to the power

$$\left[\frac{2n}{p_i}\right] + \left[\frac{2n}{p_i^2}\right] + \cdots + \left[\frac{2n}{p_i^{q_i}}\right] - 2\left\{\left[\frac{n}{p_i}\right] + \left[\frac{n}{p_i^2}\right] + \cdots + \left[\frac{n}{p_i^{q_i}}\right]\right\}$$
$$= \left(\left[\frac{2n}{p_i}\right] - 2\left[\frac{n}{p_i}\right]\right) + \left(\left[\frac{2n}{p_i^2}\right] - 2\left[\frac{n}{p_i^2}\right]\right)$$
$$+ \cdots + \left(\left[\frac{2n}{p_i^{q_i}}\right] - 2\left[\frac{n}{p_i^{q_i}}\right]\right),$$

where q_i is the largest integer such that $p_i^{q_i} \leqq 2n$. But by the definition of the integral part of a number,

$$2\left[\frac{a}{2}\right] = \begin{cases} [a] & \text{if } \dfrac{a}{2} - \left[\dfrac{a}{2}\right] < \dfrac{1}{2}, \\ [a] - 1 & \text{if } \dfrac{a}{2} - \left[\dfrac{a}{2}\right] \geqq \dfrac{1}{2}, \end{cases}$$

that is, $[a] - 2[a/2] = 0$ or 1, whatever the value of a. It follows that the sum

$$\left(\left[\frac{2n}{p_i}\right] - 2\left[\frac{n}{p_i}\right]\right) + \left(\left[\frac{2n}{p_i^2}\right] - 2\left[\frac{n}{p_i^2}\right]\right) + \cdots + \left(\left[\frac{2n}{p_i^{q_i}}\right] - 2\left[\frac{n}{p_i^{q_i}}\right]\right)$$

does not exceed $\overbrace{1 + 1 + \cdots + 1 + 1}^{q_i \text{ terms}} = q_i$. Hence p_i appears in the prime decomposition of $\binom{2n}{n}$ to a power $\leqq q_i$, and so

$$\binom{2n}{n} \leqq p_1^{q_1} p_2^{q_2} \cdots p_r^{q_r} \leqq (2n)(2n) \cdots (2n) = (2n)^r.$$

Here $r = \pi(2n)$, the number of primes not exceeding $2n$. We may therefore write the last inequality in the form

$$\binom{2n}{n} \leqq (2n)^{\pi(2n)}.$$

On the other hand,

$$\binom{2n}{n} = \frac{2n(2n-1)\cdots(n+1)}{1 \cdot 2 \cdot 3 \cdots n}$$

is divisible by the product of all the primes $p_{s+1}, p_{s+2}, \ldots, p_r$ greater than n but $\leqq 2n$. (We denote the primes $\leqq n$ by $p_1, p_2, \ldots, p_s$.) So

$$\binom{2n}{n} \geqq p_{s+1} p_{s+2} \cdots p_r.$$

On replacing each of these primes by the smaller number n, we find that

$$\binom{2n}{n} > n \cdot n \cdots n = n^{r-s},$$

where $r = \pi(2n)$ and $s = \pi(n)$. Thus we obtain the result

$$n^{\pi(2n)-\pi(n)} < \binom{2n}{n} \leqq (2n)^{\pi(2n)}. \tag{2}$$

Comparing the inequalities (1) and (2), we see that

$$2^n \leqq (2n)^{\pi(2n)}.$$

Taking logarithms, we find that

$$\pi(2n) \log (2n) \geqq n \log 2,$$

or

$$\pi(2n) \geqq \frac{\log 2}{2} \frac{2n}{\log (2n)} = 0.1501 \cdots \frac{2n}{\log (2n)}.$$

Thus for N even ($N = 2n$) we have already obtained one of the required inequalities. We can deduce a similar inequality for odd $N > 1$ by noting that $2n/(2n + 1) \geqq \frac{2}{3}$. It follows from this inequality that

$$\pi(2n + 1) \log (2n + 1) > \pi(2n) \log (2n) \geqq \frac{\log 2}{2} (2n)$$

$$\geqq \frac{2}{3} \cdot \frac{\log 2}{2} (2n + 1),$$

so that

$$\pi(2n + 1) > \frac{\log 2}{3} \frac{2n + 1}{\log (2n + 1)} = 0.10034 \cdots \frac{2n + 1}{\log (2n + 1)}.$$

Thus for all $N > 1$,

$$\pi(N) > 0.1 \frac{N}{\log N}.$$

The proof of the second of the required inequalities is somewhat more complicated. First of all, on comparing (1) and (2), we find that

$$n^{\pi(2n)-\pi(n)} < 2^{2n}.$$

Taking logarithms, we obtain $[\pi(2n) - \pi(n)] \log n < 2n \log 2$, so that

$$\pi(2n) - \pi(n) < 2 \log 2 \frac{n}{\log n} = 0.60206 \cdots \frac{n}{\log n}.$$

Suppose x is an arbitrary number >1 (not necessarily an integer). As before, we denote by $\pi(x)$ the number of primes $\leqq x$. Let $n = [x/2]$; then clearly $[x] = 2n$ or $2n + 1$, and

$$\pi(x) - \pi\left(\frac{x}{2}\right) \leqq \pi(2n) - \pi(n) + 1$$

$$< 2 \log 2 \frac{n}{\log n} + 1 < (2 \log 2 + 1) \frac{n}{\log n}$$

$$= 1.60206 \cdots \frac{n}{\log n}$$

(since $n/(\log n) > 1$). It is not hard to show that for $n \geqq 3$ and $n < x$, $n/(\log n) < x/(\log x)$. (See, for example, the solution to problem 168**a**.) It follows that for $[x/2] \geqq 3$,

$$\pi(x) - \pi\left(\frac{x}{2}\right) < (2 \log 2 + 1) \frac{x}{\log x}.$$

The last inequality also holds for $[x/2] < 3$, that is, for $x < 6$. For if $x < 10$, then $\log x < 1$, and therefore $\pi(x) - \pi(x/2) \leqq \pi(x) < x < x/(\log x) < (2 \log 2 + 1)x/(\log x)$. Thus for arbitrary $x > 1$ (integral or not) we have

$$\pi(x) - \pi\left(\frac{x}{2}\right) < (2 \log 2 + 1) \frac{x}{\log x}.$$

From this we obtain the inequality

$$\pi(x) \log x - \pi\left(\frac{x}{2}\right) \log \frac{x}{2}$$

$$= \left[\pi(x) - \pi\left(\frac{x}{2}\right)\right] \log x + \pi\left(\frac{x}{2}\right)\left(\log x - \log \frac{x}{2}\right)$$

$$< (2 \log 2 + 1) \frac{x}{\log x} \log x + \pi\left(\frac{x}{2}\right) \log 2$$

$$< \left(2 \log 2 + 1 + \frac{\log 2}{2}\right) x = 1.75257 \cdots x.$$

[Here we have used the obvious fact that $\pi(x/2) < x/2$.]

Suppose that N is an arbitrary positive integer. Then by what we have proved,

$$\pi(N) \log N - \pi\left(\frac{N}{2}\right) \log \frac{N}{2} < 1.75257 \cdots N,$$

$$\pi\left(\frac{N}{2}\right) \log \left(\frac{N}{2}\right) - \pi\left(\frac{N}{4}\right) \log \frac{N}{4} < 1.75257 \cdots \frac{N}{2},$$

$$\pi\left(\frac{N}{4}\right) \log \frac{N}{4} - \pi\left(\frac{N}{8}\right) \log \frac{N}{8} < 1.75257 \cdots \frac{N}{4},$$

$$\vdots$$

$$\pi\left(\frac{N}{2^{k-1}}\right) \log \frac{N}{2^{k-1}} - \pi\left(\frac{N}{2^k}\right) \log \frac{N}{2^k} < 1.75257 \cdots \frac{N}{2^{k-1}}.$$

Choose k so that $2^k > N$, and add all these inequalities. This gives

$$\pi(N)\log N - \pi\left(\frac{N}{2^k}\right)\log\frac{N}{2^k} < 1.75257\cdots\left(N + \frac{N}{2} + \frac{N}{4} + \cdots + \frac{N}{2^{k-1}}\right)$$

$$= 1.75257\cdots\frac{N - N/2^k}{1 - 1/2} < 3.50514\cdots N < 4N.$$

By the choice of k, $N/2^k < 1$ and therefore $\pi(N/2^k) = 0$. Thus we find that

$$\pi(N) < 4\frac{N}{\log N},$$

as required.

171. We start from the decomposition of $N! = 1 \cdot 2 \cdot 3 \cdots N$ into its prime factors:

$$N! = p_1^{\alpha_1}p_2^{\alpha_2}\cdots p_r^{\alpha_r}.$$

Here $p_1, p_2, \ldots, p_r$ are all the primes not exceeding N, and $\alpha_i (i = 1, 2, \ldots, r)$ is equal to

$$\left[\frac{N}{p_i}\right] + \left[\frac{N}{p_i^2}\right] + \left[\frac{N}{p_i^3}\right] + \cdots + \left[\frac{N}{p_i^{q_i}}\right].$$

The square brackets denote integral parts, and q_i is the largest integer such that $p_i^{q_i} \leq N$ (compare the solution to problem 170). On taking logarithms, we find that

$$\log N! = \alpha_1 \log p_1 + \alpha_2 \log p_2 + \cdots + \alpha_r \log p_r.$$

We now estimate $\log N!$ in two different ways. The right-hand side of the last equation is of the form

$$\left(\left[\frac{N}{p_1}\right] + \left[\frac{N}{p_1^2}\right] + \cdots + \left[\frac{N}{p_1^{q_i}}\right]\right)\log p_1$$
$$+ \left(\left[\frac{N}{p_2}\right] + \left[\frac{N}{p_2^2}\right] + \cdots + \left[\frac{N}{p_2^{q_2}}\right]\right)\log p_2$$
$$+ \cdots + \left(\left[\frac{N}{p_r}\right] + \left[\frac{N}{p_r^2}\right] + \cdots + \left[\frac{N}{p_r^{q_r}}\right]\right)\log p_r.$$

Let us throw away all the square brackets in this expression, that is, let us replace $[N/p_i^k]$ by N/p_i^k. In the process we introduce an error at most 1 into each term. Thus the total error introduced is at most

$$q_1 \log p_1 + q_2 \log p_2 + \cdots + q_r \log p_r$$
$$= \log p_1^{q_1} + \log p_2^{q_2} + \cdots + \log p_r^{q_r}$$
$$\leq \underbrace{\log N + \log N + \cdots + \log N}_{r \text{ times}} = r \log N.$$

By the result of the previous problem, r is less than $B(N/\log N)$, where B is a constant (in fact, we could take $B = 4$). Thus

$$r \log N < B \frac{N}{\log N} \log N = BN.$$

So we have

$$\left(\frac{N}{p_1} + \frac{N}{p_1^2} + \cdots + \frac{N}{p_1^{q_1}}\right) \log p_1 + \left(\frac{N}{p_2} + \frac{N}{p_2^2} + \cdots + \frac{N}{p_2^{q_2}}\right) \log p_2$$
$$+ \cdots + \left(\frac{N}{p_r} + \frac{N}{p_r^2} + \cdots + \frac{N}{p_r^{q_r}}\right) \log p_r \geqq \log N!$$
$$> \left(\frac{N}{p_1} + \frac{N}{p_1^2} + \cdots + \frac{N}{p_1^{q_1}}\right) \log p_1 + \left(\frac{N}{p_2} + \frac{N}{p_2^2} + \cdots + \frac{N}{p_2^{q_2}}\right) \log p_2$$
$$+ \cdots + \left(\frac{N}{p_r} + \frac{N}{p_r^2} + \cdots + \frac{N}{p_r^{q_r}}\right) \log p_r - BN.$$

We now use the inequalities

$$\frac{N}{p_i} + \frac{N}{p_i^2} + \cdots + \frac{N}{p_i^{q_i}} \geqq \frac{N}{p_i}$$

and

$$\frac{N}{p_i} + \frac{N}{p_i^2} + \cdots + \frac{N}{p_i^{q_i}} = \frac{(N/p_i) - (N/p_i^{q_i+1})}{1 - (1/p_i)} < \frac{N/p_i}{1 - (1/p_i)} = \frac{N}{p_i - 1}.$$

Thus we find that

$$\frac{N}{p_i - 1} \log p_1 + \frac{N}{p_2 - 1} \log p_2 + \cdots + \frac{N}{p_r - 1} \log p_r > \log N!$$
$$> \frac{N}{p_2} \log p_1 + \frac{N}{p_2} \log p_2 + \cdots + \frac{N}{p_r} \log p_r - BN. \quad (1)$$

From this one can deduce the inequalities

$$N\left\{\frac{\log p_1}{p_1} + \frac{\log p_2}{p_2} + \cdots + \frac{\log p_r}{p_r} + K\right\} > \log N!$$
$$> N\left\{\frac{\log p_1}{p_1} + \frac{\log p_2}{p_2} + \cdots + \frac{\log p_r}{p_r} - B\right\}, \quad (2)$$

where K is a positive constant. For

$$\frac{\log p_i}{p_i - 1} - \frac{\log p_i}{p_i} = \frac{\log p_i}{p_i(p_i - 1)} \qquad (i = 1, 2, \ldots, r),$$

so that the left-hand side of (1) differs from

$$N\left(\frac{\log p_1}{p_1} + \frac{\log p_2}{p_2} + \cdots + \frac{\log p_r}{p_r}\right)$$

by the quantity

$$N\left\{\frac{\log p_i}{p_i(p_i-1)}+\frac{\log p_2}{p_2(p_2-1)}+\cdots+\frac{\log p_r}{p_r(p_r-1)}\right\}.$$

We show that for all r the sum inside the braces is bounded. Note that for each integer $a \geqq 2$

$$\frac{\log a}{a(a-1)}<\frac{1}{a\sqrt{a}}, \quad \text{that is,} \quad \log a<\frac{a-1}{\sqrt{a}}.$$

For if $a \geqq 2$, then

$$2\left(\frac{a-1}{\sqrt{a}}\right)=2\left(\sqrt{a}-\frac{\sqrt{a}}{a}\right) \geqq 2\left(\sqrt{a}-\frac{\sqrt{a}}{2}\right)=\sqrt{a},$$

while

$$2\log a<\sqrt{a}, \quad \text{since} \quad 10^{2\log a}=a^2<10^{\sqrt{a}}$$

To see why this last inequality holds, note that if a^2 lies between 10^k and 10^{k+1}, then the number of digits in $10^{\sqrt{a}}$ is the integral part of $\sqrt{a}$ plus 1. This is clearly greater than $k+1$. It follows from this inequality that

$$\frac{\log p_1}{p_1(p_1-1)}+\frac{\log p_2}{p_2(p_2-1)}+\cdots+\frac{\log p_r}{p_r(p_r-1)}$$
$$<\frac{1}{p_1\sqrt{p_1}}+\frac{1}{p_2\sqrt{p_2}}+\cdots+\frac{1}{p_r\sqrt{p_r}}$$
$$<1+\frac{1}{2\sqrt{2}}+\frac{1}{3\sqrt{3}}+\frac{1}{4\sqrt{4}}+\cdots\frac{1}{N\sqrt{N}}.$$

But this last sum remains bounded for all N:

$$1+\frac{1}{2\sqrt{2}}+\frac{1}{3\sqrt{3}}+\frac{1}{4\sqrt{4}}+\cdots+\frac{1}{N\sqrt{N}}<\frac{3/2}{3/2-1}=3$$

(see problem 169). Thus in the inequality (2), K can be taken equal to 3.

In order to obtain another estimate for $\log N!$ we use the fact that for any N

$$C_1\sqrt{N}\left(\frac{N}{e}\right)^N>N!>C_2\sqrt{N}\left(\frac{N}{e}\right)^N,$$

where $C_1=e$, $C_2=\sqrt{\tfrac{4}{5}}\,e$ (see problem 165). On taking logarithms we find that

$$N\{\log N-\log e\}+\frac{1}{2}\log N+\log C_1>\log N!$$
$$>N\{\log N-\log e\}+\frac{1}{2}\log N+\log C_2.$$

We use the facts that $\frac{1}{2}(\log N)/N \leqq \frac{1}{2}(\log 3)/3 < 0.08$ for $N \geqq 3$ (see the solution to 168a) and that $(\log C_1)/N \leqq (\log e)/3 < 0.15$, to write this inequality in the form

$$N\{\log N - \log e + 0.23\} > \log N! > N\{\log N - \log e\}.$$

On comparing this with the inequality (2) on p. 184, we find that

$$\frac{\log p_1}{p_1} + \frac{\log p_2}{p_2} + \cdots + \frac{\log p_r}{p_r} + K < \log N - \log e,$$

$$\frac{\log p_1}{p_1} + \frac{\log p_2}{p_2} + \cdots + \frac{\log p_r}{p_r} - B < \log N - \log e + 0.23,$$

whence we obtain the required result:

$$\log N + R > \frac{\log p_1}{p_1} + \frac{\log p_2}{p_2} + \cdots + \frac{\log p_r}{p_r} > \log N - R,$$

where we can take for R the larger of the numbers $B - \log e + 0.23$ and $K + \log e$. Since the constants B and K can be given the values 4 and 3, respectively, and since $\log e \approx 0.4343$, we can take $R = 4$.

172a. From the definition of $B_1, B_2, \ldots, B_n$ we have $b_1 = B_1, b_2 = B_2 - B_1$, $b_3 = B_3 - B_2, \ldots, b_n = B_n - B_{n-1}$. Thus

$$a_1b_1 + a_2b_2 + a_3b_3 + \cdots + a_nb_n$$
$$= a_1b_1 + a_2(B_2 - B_1) + a_3(B_3 - B_2) + \cdots + a_n(B_n - B_{n-1}),$$

or, regrouping terms,

$$a_1b_1 + a_2b_2 + \cdots + a_nb_n$$
$$= (a_1 - a_2)B_1 + (a_2 - a_3)B_2 + \cdots + (a_{n-1} - a_n)B_{n-1} + a_nB_n,$$

as required.

b. (1) If we substitute

$$a_1 = 1, a_2 = 2, a_3 = 3, \ldots, a_n = n;$$
$$b_1 = 1, b_2 = q\ \ b_3 = q^2, \ldots, b_n = q^{n-1}$$

in the formula of part **a**, we find that

$$B_1 = 1\left(= \frac{q-1}{q-1}\right), \qquad B_2 = 1 + q = \frac{q^2-1}{q-1},$$

$$B_3 = 1 + q + q^2 = \frac{q^3-1}{q-1}, \ldots,$$

$$B_{n-1} = 1 + q + q^2 + \cdots + q^{n-2} = \frac{q^{n-1}-1}{q-1},$$

$$B_n = 1 + q + q^2 + \cdots + q^{n-1} = \frac{q^n-1}{q-1},$$

and

$$a_1 - a_2 = a_2 - a_3 = a_3 - a_4 = \cdots = a_{n-1} - a_n = -1.$$

Thus we find that

$$\begin{aligned}
&1 + 2q + 3q^2 + \cdots + nq^{n-1} \\
&\quad = -\left(\frac{q-1}{q-1} + \frac{q^2-1}{q-1} + \cdots + \frac{q^{n-1}-1}{q-1}\right) + n\frac{q^n-1}{q-1} \\
&\quad = \frac{-1}{q-1}\left(\frac{q^n-1}{q-1} - n\right) + n\frac{q^n-1}{q-1} = \frac{nq^n}{q-1} - \frac{q^n-1}{(q-1)^2}.
\end{aligned}$$

(2) If we substitute

$$a_1 = 1, a_2 = 4, a_3 = 9, \ldots, a_n = n^2;$$
$$b_1 = 1, b_2 = q, b_3 = q^2, \ldots, b_n = q^{n-1}$$

in the formula of part **a**, then $B_1, B_2, \ldots, B_n$ have the same values as before, and

$$a_1 - a_2 = -3, a_2 - a_3 = -5, a_3 - a_4 = -7, \ldots,$$
$$a_{n-1} - a_n = (n-1)^2 - n^2 = -(2n-1).$$

Thus we find that

$$\begin{aligned}
&1 + 4q + 9q^2 + \cdots + n^2q^{n-1} \\
&\quad = -\left(3\frac{q-1}{q-1} + 5\frac{q^2-1}{q-1} + 7\frac{q^3-1}{q-1}\right. \\
&\qquad \left. + \cdots + (2n-1)\frac{q^{n-1}-1}{q-1}\right) + n^2\frac{q^n-1}{q-1} \\
&\quad = \left(\frac{q-1}{q-1} + \frac{q^2-1}{q-1} + \frac{q^3-1}{q-1} + \cdots + \frac{q^{n-1}-1}{q-1}\right) \\
&\qquad - 2\left(2\frac{q-1}{q-1} + 3\frac{q^2-1}{q-1} + \cdots + n\frac{q^{n-1}-1}{q-1}\right) + n^2\frac{q^n-1}{q-1} \\
&\quad = \frac{1}{q-1}(1 + q + q^2 + \cdots + q^{n-1} - n) \\
&\qquad - \frac{2}{q-1}\left(1 + 2q + 3q^2 + \cdots + nq^{n-1} - \frac{n(n+1)}{2}\right) + n^2\frac{q^n-1}{q-1} \\
&\quad = \frac{1}{q-1}\left(\frac{q^n-1}{q-1} - n\right) \\
&\qquad - \frac{2}{q-1}\left[\frac{nq^n}{q-1} - \frac{q^n-1}{(q-1)^2} - \frac{n(n+1)}{2}\right] + n^2\cdot\frac{q^n-1}{q-1} \\
&\quad = \frac{n^2q^n}{q-1} - \frac{(2n-1)q^n+1}{(q-1)^2} + \frac{2q^n-2}{(q-1)^3}.
\end{aligned}$$

Here we have used the fact that by part 1,

$$1 + 2q + 3q^2 + \cdots + nq^{n-1} = \frac{nq^n}{q-1} - \frac{q^n - 1}{(q-1)^2}.$$

173a. In order to evaluate

$$\frac{1}{p_1} + \frac{1}{p_2} + \frac{1}{p_3} + \cdots + \frac{1}{p_n},$$

where $p_1, p_2, \ldots, p_n$ are all the prime numbers not exceeding some given integer N, we substitute

$$a_1 = \frac{1}{\log p_1}, \quad a_2 = \frac{1}{\log p_2}, \quad a_3 = \frac{1}{\log p_3}, \ldots, a_r = \frac{1}{\log p_r},$$

$$b_1 = \frac{\log p_1}{p_1}, \quad b_2 = \frac{\log p_2}{p_2}, \quad b_3 = \frac{\log p_3}{p_3}, \ldots, b_r = \frac{\log p_r}{p_r}$$

in the formula of problem 172a.

Put

$$\frac{\log p_1}{p_1} = B_1$$

$$\frac{\log p_2}{p_2} + \frac{\log p_1}{p_1} = B_2$$

$$\vdots$$

$$\frac{\log p_r}{p_r} + \cdots + \frac{\log p_2}{p_2} + \frac{\log p_1}{p_1} = B_r.$$

We find that

$$\frac{1}{p_1} + \frac{1}{p_2} + \frac{1}{p_3} + \cdots + \frac{1}{p_r}$$

$$= \left(\frac{1}{\log p_1} - \frac{1}{\log p_2}\right)B_1 + \left(\frac{1}{\log p_2} - \frac{1}{\log p_3}\right)B_2$$

$$+ \left(\frac{1}{\log p_3} - \frac{1}{\log p_4}\right)B_3 + \cdots + \left(\frac{1}{\log p_{r-1}} - \frac{1}{\log p_r}\right)B_{r-1} + \frac{1}{\log p_r}B_r$$

$$= \left(\frac{1}{\log p_1} - \frac{1}{\log p_2}\right)B_1 + \left(\frac{1}{\log p_2} - \frac{1}{\log p_3}\right)B_2$$

$$+ \left(\frac{1}{\log p_3} - \frac{1}{\log p_4}\right)B_3 + \cdots + \left(\frac{1}{\log p_{r-1}} - \frac{1}{\log p_r}\right)B_{r-1}$$

$$+ \left(\frac{1}{\log p_r} - \frac{1}{\log N}\right)B_r + \frac{1}{\log N}B_r.$$

Let us now estimate this last expression. We first use the fact that by Mertens' first formula (problem 171),

$$B_1 < \log p_1 + R,$$
$$B_2 < \log p_2 + R,$$
$$\vdots$$
$$B_{r-1} < \log p_{r-1} + R$$
$$B_r < \log p_r + R \quad \text{and} \quad B_r < \log N + R.$$

Here R is a constant (which can be taken equal to 4). From this it follows that

$$\begin{aligned}
\frac{1}{p_1} + \frac{1}{p_2} + \frac{1}{p_3} + \cdots + \frac{1}{p_r} &< \left(\frac{1}{\log p_1} - \frac{1}{\log p_2}\right)(\log p_1 + R) \\
&\quad + \left(\frac{1}{\log p_2} - \frac{1}{\log p_3}\right)(\log p_2 + R) \\
&\quad + \left(\frac{1}{\log p_3} - \frac{1}{\log p_4}\right)(\log p_3 + R) \\
&\quad + \cdots + \left(\frac{1}{\log p_{r-1}} - \frac{1}{\log p_r}\right) \cdot (\log p_{r-1} + R) \\
&\quad + \left(\frac{1}{\log p_r} - \frac{1}{\log N}\right)(\log p_r + R) + \frac{1}{\log N}(\log N + R) \\
&= 1 + \Bigg[-\frac{1}{\log p_2}\log p_1 + \left(\frac{1}{\log p_2} - \frac{1}{\log p_3}\right)\log p_2 \\
&\quad + \left(\frac{1}{\log p_3} - \frac{1}{\log p_4}\right)\log p_3 + \cdots + \left(\frac{1}{\log p_{r-1}} - \frac{1}{\log p_r}\right)\log p_{r-1} \\
&\quad + \left(\frac{1}{\log p_r} - \frac{1}{\log N}\right)\log p_r + \frac{1}{\log N}\log N\Bigg] + R\frac{1}{\log p_1},
\end{aligned}$$

since the terms having R as a factor cancel out in pairs (except for $R/\log p_1$).

It remains only to estimate the sum inside the brackets. This may easily be done geometrically. Write the sum in the form

$$\begin{aligned}
&(\log p_2 - \log p_1)\frac{1}{\log p_2} + (\log p_3 - \log p_2)\frac{1}{\log p_3} \\
&\quad + (\log p_4 - \log p_3)\frac{1}{\log p_4} + \cdots \\
&\quad + (\log p_r - \log p_{r-1})\frac{1}{\log p_r} + (\log N - \log p_r)\frac{1}{\log N}.
\end{aligned}$$

Now look at fig. 116a, in which the hyperbola $y = 1/x$ is shown. It is clear that the expression in which we are interested is the area of the shaded part of the figure and is therefore less than the area bounded by the hyperbola, the x axis, and the lines $x = \log p_1$ and $x = \log N$. But this area is $\ln(\log N) - \ln(\log p_1)$ (see problem 154).

Thus we obtain, finally, the inequality

$$\frac{1}{p_1} + \frac{1}{p_2} + \frac{1}{p_3} + \cdots + \frac{1}{p_r} < \ln \log N - \ln \log p_1 + R\frac{1}{\log p_1} + 1. \quad (1)$$

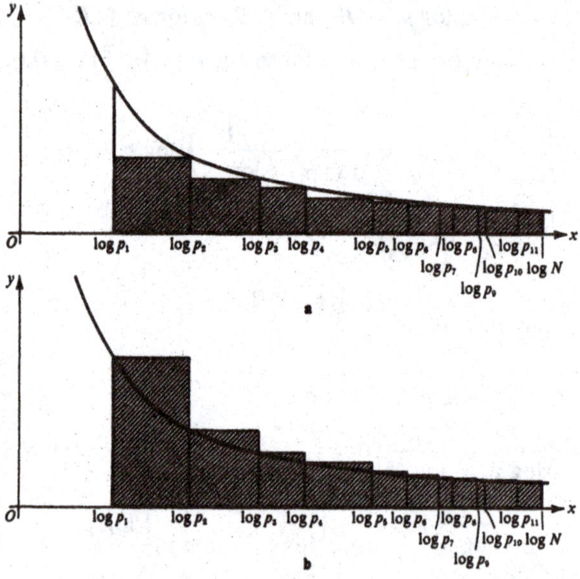

Fig. 116

We have now found an upper bound for our sum. To obtain a lower bound, we start by transforming Mertens' first formula somewhat. It follows from the formula that

$$\begin{aligned} B_i &= \frac{\log p_1}{p_1} + \frac{\log p_2}{p_2} + \cdots + \frac{\log p_i}{p_i} \\ &> \frac{\log p_1}{p_1} + \frac{\log p_2}{p_2} + \cdots + \frac{\log p_i}{p_i} + \frac{\log p_{i+1}}{p_{i+1}} - a \\ &> \log p_{i+1} - R - a \qquad (i = 1, 2, \ldots, r), \end{aligned}$$

where a is a constant chosen greater than $(\log p_{i+1})/p_{i+1}$ $(i = 1, 2, \ldots, r)$. (We could, for example, take $a = (\log 3)/3$ or $a = 0.16 > (\log 3)/3$; see the solution to problem 171.) We also use the fact that

$$\frac{\log p_1}{p_1} + \frac{\log p_2}{p_2} + \frac{\log p_3}{p_3} + \cdots + \frac{\log p_r}{p_r} > \log N - R - a.$$

(According to Mertens' first formula the summand $-a$ is actually unnecessary, but we shall find it convenient to include it.) We obtain

$$
\begin{aligned}
\frac{1}{p_1}+\frac{1}{p_2}+\frac{1}{p_3}+\cdots+\frac{1}{p_n} &> \left(\frac{1}{\log p_1}-\frac{1}{\log p_2}\right)(\log p_2 - R - a)\\
&\quad+\left(\frac{1}{\log p_2}-\frac{1}{\log p_3}\right)(\log p_3 - R - a)\\
&\quad+\left(\frac{1}{\log p_3}-\frac{1}{\log p_4}\right)(\log R_4 - R - a)\\
&\quad+\cdots+\left(\frac{1}{\log p_{r-1}}-\frac{1}{\log p_r}\right)(\log p_r - R - a)\\
&\quad+\left(\frac{1}{\log p_r}-\frac{1}{\log N}\right)(\log N - R - a)\\
&\quad+\frac{1}{\log N}(\log N - R - a)\\
&= \left[(\log p_2 - \log p_1)\frac{1}{\log p_1} + (\log p_3 - \log p_2)\frac{1}{\log p_2}\right.\\
&\quad+(\log p_4 - \log p_3)\frac{1}{\log p_3}+\cdots+(\log p_r - \log p_{r-1})\frac{1}{\log p_{r-1}}\\
&\quad\left.+(\log N - \log p_r)\frac{1}{\log p_r}\right] + 1 - (R+a)\frac{1}{\log p_1}.
\end{aligned}
$$

In the last expression we have omitted the term $-(1/\log N)\log N$ from the quantity inside the brackets and included the extra term $-(1/\log p_1)\log p_1$. It is easy to see that the sum inside the brackets is greater than the area bounded by the hyperbola $y = 1/x$, the x axis, and the lines $x = \log N$ and $x = \log p_1$. (See fig. 116b, in which the shaded area is equal to this sum.) Thus

$$
\frac{1}{p_1}+\frac{1}{p_2}+\frac{1}{p_3}+\cdots+\frac{1}{p_r} > \ln\log N - \ln\log p_1 - (R+a)\frac{1}{\log p_1} + 1. \tag{2}
$$

To avoid using logarithms to two different bases in the same formula, let us change to natural logarithms throughout. To transform common logarithms to natural logarithms, we use the formula

$$\log N = M \ln N,$$

where $M = \log e = 0.434\ldots.$ (See footnote on page 159.)

Thus our estimates (1) and (2) for the sum

$$S = \frac{1}{p_1}+\frac{1}{p_2}+\frac{1}{p_3}+\cdots+\frac{1}{p_r}$$

assume the form

$$\ln \ln N + \ln M - \ln \log 2 + \frac{R}{\log 2} + 1$$
$$> \frac{1}{p_1} + \frac{1}{p_2} + \frac{1}{p_3} + \cdots + \frac{1}{p_r}$$
$$> \ln \ln N + \ln M - \ln \log 2 - \frac{R + a}{\log 2} + 1$$

(recalling that p_1 is the first prime, so that $p_1 = 2$). It follows from this double inequality that there is a constant T such that S lies between $\ln \ln N + T$ and $\ln \ln N - T$. For T we may take the larger of the numbers $R/(\log 2) + 1 + \ln M - \ln \log 2$ and $(R + a)/(\log 2) - 1 - \ln M + \ln \log 2$; since the constants $R/(\log 2)$, $(R + a)/(\log 2)$, $\ln M$ and $\ln \log 2$ may be assigned the values $4/0.301 \cdots < 13\frac{1}{2}$, $4.16/0.301 \cdots < 14$, $\ln 0.434 \cdots = -0.833 \cdots$ and $\ln 0.301 \cdots = -1.20 \ldots$, respectively, we can take $T = 15$.

b. Let us use natural logarithms from the beginning. We write

$$B_1 = \frac{\ln p_1}{p_1}, \qquad B_2 = \frac{\ln p_1}{p_1} + \frac{\ln p_2}{p_2}, \ldots,$$
$$B_r = \frac{\ln p_1}{p_1} + \frac{\ln p_2}{p_2} + \cdots + \frac{\ln p_r}{p_r},$$

and introduce the differences

$$\alpha_1 = B_1 - \ln p_1, \alpha_2 = B_2 - \ln p_2, \ldots, \alpha_r = B_r - \ln p_r.$$

Let us also write

$$\alpha^{(N)} = B_r - \ln N,$$

where p_r is the largest prime $\leq N$.

It follows from Mertens' first formula (problem 171) that all these differences are bounded. In absolute value they are all less than $4/0.434 \cdots < 10$.[20] Using the result of problem 172 in the same way as in the solution to part **a**, we find that

$$\varepsilon^{(N)} = \frac{1}{p_1} + \frac{1}{p_2} + \cdots + \frac{1}{p_n} - \ln \ln N = \varepsilon_1^{(N)} + \varepsilon_2^{(N)},$$

[20] Since

$$\frac{\ln p_1}{p_1} + \frac{\ln p_2}{p_2} + \cdots + \frac{\ln p_k}{p_k} - \ln p_k = \frac{1}{M}\left(\frac{\log p_1}{p_1} + \frac{\log p_2}{p_2} + \cdots + \frac{\log p_k}{p_k} - \log p_k\right),$$

where $M = 0.434 \cdots$ (see the footnote on p. 159), and

$$\left|\frac{\log p_1}{p_1} + \frac{\log p_2}{p_2} + \cdots + \frac{\log p_k}{p_k} - \log p_k\right| < 4.$$

where

$$\varepsilon_1^{(N)} = \alpha_1\left(\frac{1}{\ln p_1} - \frac{1}{\ln p_2}\right) + \alpha_2\left(\frac{1}{\ln p_2} - \frac{1}{\ln p_3}\right)$$
$$+ \cdots + \alpha_{r-1}\left(\frac{1}{\ln p_{r-1}} - \frac{1}{\ln p_r}\right) + \alpha_r\left(\frac{1}{\ln p_r} - \frac{1}{\ln N}\right) + \alpha^{(N)}\frac{1}{\ln N},$$

and

$$\varepsilon_2^{(N)} = \ln p_1\left(\frac{1}{\ln p_1} - \frac{1}{\ln p_2}\right) + \ln p_2\left(\frac{1}{\ln p_2} - \frac{1}{\ln p_3}\right)$$
$$+ \cdots + \ln p_{r-1}\left(\frac{1}{\ln p_{r-1}} - \frac{1}{\ln p_r}\right) + \ln p_r\left(\frac{1}{\ln p_r} - \frac{1}{\ln N}\right)$$
$$+ \ln N\frac{1}{\ln N} - \ln\ln N$$
$$= 1 + (\ln p_2 - \ln p_1)\frac{1}{\ln p_2} + (\ln p_3 - \ln p_2)\frac{1}{\ln p_3}$$
$$+ \cdots + (\ln p_r - \ln p_{r-1})\frac{1}{\ln p_r} + (\ln N - \ln p_r)\frac{1}{\ln N} - \ln\ln N.$$

In the solution to part a we showed that $|\varepsilon^{(N)}| < 15$ for any value of N. Here we must show that as $N \to \infty$ the sequence $\{\varepsilon^{(N)}\} = \{\varepsilon_1^{(N)} + \varepsilon_2^{(N)}\}$ tends to a limit. To do this it is sufficient to show that each of the sequences $\{\varepsilon_1^{(N)}\}$ and $\{\varepsilon_2^{(N)}\}$ tends to a limit as $N \to \infty$.

As far as the sequence $\{\varepsilon_2^{(N)}\}$ is concerned, the proof is very simple. We need only use the fact that

$$-\varepsilon_2^{(N)} - \ln\ln 2 + 1 = \ln\frac{\ln N}{\ln 2} - \left[(\ln p_2 - \ln p_1)\frac{1}{\ln p_2}\right.$$
$$+ (\ln p_3 - \ln p_2)\frac{1}{\ln p_3} + \cdots + (\ln p_r - \ln p_{r-1})$$
$$\left.\times\frac{1}{\ln p_r} + (\ln N - \ln p_r)\frac{1}{\ln N}\right]$$

is equal to the sum of the areas of the curvilinear triangles not shaded in fig. 116a. We see from this that $-\varepsilon_2^{(N)} - \ln\ln 2 + 1$ increases monotonically as N increases; and since the terms of the sequence remain bounded above, it must have a limit as $N \to \infty$. And since $-\varepsilon_2^{(N)} - \ln\ln 2 + 1$ tends to a limit, so does $\varepsilon_2^{(N)}$.

We show now that the sequence $\varepsilon_1^{(N)}$ also tends to some limit as $N \to \infty$. Let us write

$$\alpha_1\left(\frac{1}{\ln p_1} - \frac{1}{\ln p_2}\right) + \alpha_2\left(\frac{1}{\ln p_2} - \frac{1}{\ln p_3}\right) + \cdots$$
$$+ \alpha_{k-1}\left(\frac{1}{\ln p_{k-1}} - \frac{1}{\ln p_k}\right) = d_k.$$

Then it is clear that if $N > p_k$,

$$\varepsilon_1^{(N)} = d_k + \alpha_k\left(\frac{1}{\ln p_k} - \frac{1}{\ln p_{k+1}}\right) + \alpha_{k+1}\left(\frac{1}{\ln p_{k+1}} - \frac{1}{\ln p_{k+2}}\right) + \cdots$$

$$+ \alpha_{r-1}\left(\frac{1}{\ln p_{r-1}} - \frac{1}{\ln p_r}\right) + \alpha_r\left(\frac{1}{\ln p_r} - \frac{1}{\ln N}\right) + \alpha^{(N)}\frac{1}{\ln N}.$$

But all the terms $\alpha_1, \alpha_2, \ldots, \alpha_r$; $\alpha^{(N)}$ lie between -10 and $+10$. On substituting -10 and $+10$ for the α_i in the appropriate parts of the last formula, we find that

$$d_k - \frac{10}{\ln p_k} < \varepsilon_1^{(N)} < d_k + \frac{10}{\ln p_k}.$$

Thus all the numbers $\varepsilon_1^{(N)}$ for which $N > p_k$ are included within a segment of the real line of length $20/(\ln p_k)$. On choosing $l > k$ we can find a

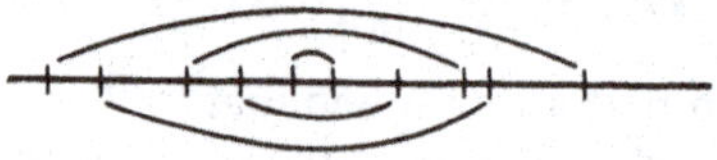

Fig. 117

segment of still smaller length within which all the $\varepsilon_1^{(N)}$ with $N > p_l$ are confined; then we choose a segment of length $20/(\ln p_m)$, $m > l$, which is smaller still, and within which all the $\varepsilon_1^{(N)}$ with $N > p_m$ lie, and so on. We obtain a system of nested intervals of length tending to zero (fig. 117). The left-hand endpoints of these intervals form an increasing sequence of numbers, which remains bounded; the right-hand endpoints form a decreasing sequence which is bounded below. Thus these sequences tend to limits $\bar{\varepsilon}_1$ and $\bar{\bar{\varepsilon}}_1$. Now there exist intervals of arbitrarily small length in our system, and the limit points must both lie inside all of them. It follows that the two limit points must coincide: $\bar{\varepsilon}_1 = \bar{\bar{\varepsilon}}_1 = \varepsilon_1$. It is easy to see that ε_1 is also the limit of the sequence $\{\varepsilon_1^{(N)}\}$.

174. We need to evaluate the product

$$\Pi^{(N)} = \left(1 - \frac{1}{p_1}\right)\left(1 - \frac{1}{p_2}\right)\left(1 - \frac{1}{p_3}\right)\cdots\left(1 - \frac{1}{p_r}\right),$$

where $p_1, p_2, \ldots, p_r$ are all the prime numbers not exceeding the integer N. Taking natural logarithms, we find

$$\ln \Pi^{(N)} = \ln\left(1 - \frac{1}{p_1}\right) + \ln\left(1 - \frac{1}{p_2}\right) + \cdots + \ln\left(1 - \frac{1}{p_r}\right).$$

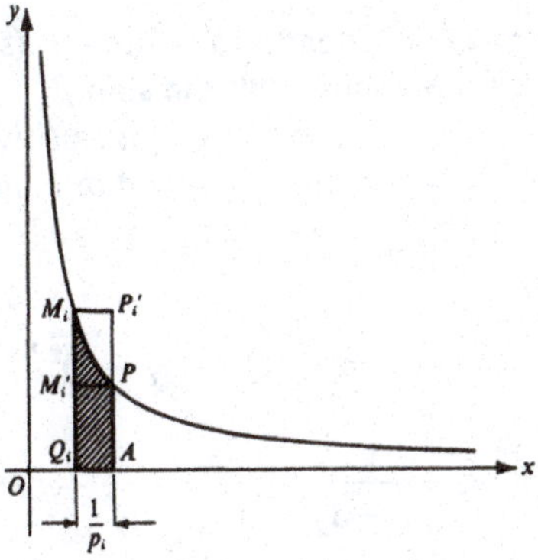

Fig. 118

The natural logarithm $\ln(1 - (1/p_i))$ is equal to the area of the curvilinear trapezoid APM_iQ_i taken with a negative sign (fig. 118). It is bounded by the x axis, the hyperbola $y = 1/x$, and the lines $x = 1 - (1/p_i)$ and $x = 1$ (see the definition of the function F on p. 30 and problem 154). But the area of the curvilinear trapezoid APM_iQ_i lies between the areas of the rectangles $APM_i'Q_i$ and $AP_i'M_iQ_i$. Now

$$S(APM_i'Q_i) = AP \cdot Q_iA = 1 \cdot \frac{1}{p_i} = \frac{1}{p_i},$$

and

$$S(AP_iM_iQ_i) = Q_iM_iQ_iA = \frac{1}{1-(1/p_i)} \cdot \frac{1}{p_i} = \frac{1}{p_i - 1},$$

so that

$$\frac{1}{p_i} < -\ln\left(1 - \frac{1}{p_i}\right) < \frac{1}{p_i - 1}.$$

Let us write

$$-\ln\left(1 - \frac{1}{p_i}\right) = \frac{1}{p_i} + \beta_i,$$

where

$$0 < \beta_i < \frac{1}{p_i - 1} - \frac{1}{p_i} = \frac{1}{p_i(p_i - 1)}.$$

The formula for $\ln \Pi^{(N)}$ now assumes the form

$$\ln \Pi^{(N)} = -\left(\frac{1}{p_1} + \frac{1}{p_2} + \cdots + \frac{1}{p_r}\right) - (\beta_1 + \beta_2 + \cdots + \beta_r),$$

or, if we write $1/p_1 + 1/p_2 + \cdots + 1/p_r - \ln \ln N = \varepsilon^{(N)}$ (see 173**b**),

$$\ln \Pi^{(N)} = -\ln \ln N - \varepsilon^{(N)} - (\beta_1 + \beta_2 + \cdots + \beta_r).$$

As $N \to \infty$, the number $\varepsilon^{(N)}$ tends to a limit ε (see Mertens' second formula, problem 173). We show that the sum $\beta_1 + \beta_2 + \cdots + \beta_r$ also tends to a limit as $N \to \infty$. Since every β_i is positive, the sums increase with r. Therefore, to show that the sums tend to a limit, it is sufficient to show that they are bounded above. But

$$\beta_i < \frac{1}{p_i(p_i - 1)} < \frac{1}{(p_i - 1)^2},$$

so that

$$\beta_1 + \beta_2 + \cdots + \beta_r < \frac{1}{(p_1 - 1)^2} + \frac{1}{(p_2 - 1)^2} + \cdots + \frac{1}{(p_r - 1)^2}$$

$$< \frac{1}{1^2} + \frac{1}{2^2} + \cdots + \frac{1}{N^2} < 2.$$

(See problem 169.) So $\beta_1 + \beta_2 + \cdots + \beta_r$ does tend to a limit as $N \to \infty$.

Thus as $N \to \infty$, the sum

$$\ln \Pi^{(N)} + \ln \ln N = -\varepsilon^{(N)} - (\beta_1 + \beta_2 + \cdots + \beta_r)$$

tends to a limit a:

$$\ln \Pi^{(N)} + \ln \ln N \to a.$$

On eliminating the logarithms, we find that

$$\Pi^{(N)} \ln N \to e^a,$$

from which the required result follows on writing c for e^a:

$$\Pi^{(N)} \sim c/\ln N.$$

HINTS AND ANSWERS

HINTS AND ANSWERS

101. Yes.

102. A network can be constructed which contains seven lines and satisfies the conditions of the problem.

103a. Select a line a, and suppose it contains $n + 1$ stops. Show that (1) there are exactly $n + 1$ lines through every stop not lying on a; (2) every line contains exactly $n + 1$ stops; (3) there are exactly $n + 1$ lines through each stop on the line a.

b. Use the result of part a.

104b. Show first that *every* point of intersection of two of the seven lines must be one of the seven points (and that every line joining two of the seven points must be one of the given lines).

105. The proof is by contradiction: suppose that the conclusion is false and that outside one of the n lines (say MN) there exist other points of intersection of the lines. Show that, given any one such point of intersection, it is always possible to find another, lying closer to MN. This will imply that there are an infinity of points of intersection, and this clearly cannot happen.

106. If the points are not all collinear, consider a triangle T with minimal altitude formed by them. By hypothesis there are at least three points of S on the line containing the base of T. Using this fact, derive a contradiction by showing that there is a triangle with an altitude smaller than that of T.

107. Use mathematical induction, making use of the result of problem 106.

108. There are six possible configurations consisting of four points which satisfy the conditions of the problem. These are

(1) The vertices of a rhombus, one of whose diagonals has the same length as a side.

(2) and (3) The vertices of a deltoid (a quadrilateral having two pairs of adjacent equal sides), the two diagonals of which are equal in length to one of the pairs of sides. The deltoid may be convex or concave.

(4) The vertices of a square.
(5) The three vertices and the center of an equilateral triangle.
(6) The vertices of an isosceles trapezoid whose sides are equal in length to the shorter base and whose diagonals are equal to the longer.

The possible values of the ratio b/a are

$$\sqrt{3}, \sqrt{2+\sqrt{3}}, \sqrt{2} \quad \text{and} \quad \tfrac{1}{2}(1+\sqrt{5}).$$

(Here b is the greater of the two distances.)

b. The possible values of n are

$n = 3$: the vertices of an isosceles triangle;
$n = 4$: the six configurations of part a;
$n = 5$: the vertices of a regular pentagon.

109a. One can give many examples of configurations of N points satisfying the conditions of the problem. In finding some of these, it is convenient to use the fact that if u and v are positive integers, then

$$x = 2uv, \quad y = |u^2 - v^2|, \quad z = u^2 + v^2$$

are the sides of a right triangle.

b. Show that if O, P, Q are three noncollinear points of the plane, then there are at most two possible positions for a point A such that the differences $AP - AO$ and $AQ - AO$ have preassigned values. It follows that there exist only a finite number of points whose distances from O, P, and Q are all integral.

110. The lattice squares decompose M into pieces. Translate these pieces so that they all lie in one lattice square and use the fact that their total area is >1.

111a. Starting with the given parallelogram, construct an infinite network of congruent parallelograms, including the given one, and covering the whole plane without overlapping. Then show that the parallelograms of this network can be divided into pieces which can be rearranged to form the squares of the original network.

b. Show that if a polygon with vertices at lattice points is divided into two smaller polygons, whose vertices are also lattice points and for which the formula holds, then it holds for the large polygon as well. Then divide the given polygon into triangles which contain no lattice points except their vertices. By the result of part **a**, the formula holds for each of them.

112. Shrink the set K to half its linear dimensions by applying the transformation $(x,y) \to (x/2,y/2)$. Apply problem 110 to the set K' thus obtained.

113. Suppose the radius ρ of the trees is greater than $\frac{1}{50}$. Draw through the center of the orchard an arbitrary straight line cutting the boundary in M and N, and construct a rectangle of width 2ρ having MN as the longer midline. By Minkowski's theorem (see problem 112) it follows that within this rectangle there are at least two points at which trees are planted, and these trees will block the view from the center in each of the directions OM and ON.

If the radius of all the trees is less than $1/\sqrt{2501}$, then we can actually construct a ray through O not blocked by a tree.

114. Use mathematical induction.

115. Start coloring the lines in accordance with the conditions of the problem. In part a, no trouble can ever arise. In part b, we may come to a line l whose four neighbors have already been colored, each with a different color. Show that in that case we can recolor some of the lines so as to release a color for l.

116. It is not hard to show that there are an odd number of 12-segments on the 12-side of T: on the remaining sides of T there are no such segments. Next, count the number of 12-sides of the small triangles. The number of such sides which belong to 123-triangles has the same parity as the number of 12-sides on a side of T. Since the latter is odd, so also is the former (and therefore certainly not zero).

b. The theorem is formulated as follows:

Suppose a tetrahedron T, with vertices marked 1, 2, 3, 4, is divided into a number of smaller tetrahedra in such a way that any two of the small tetrahedra do not touch at all, or touch at a common vertex, or along a common edge (but not part of an edge), or on a common face (but not part of a face). Suppose further that all the vertices of the small tetrahedra are numbered 1, 2, 3, or 4, in such a way that every vertex on the ijk-face of T is numbered i, j, or k, and every vertex on the ij-edge of T is numbered i or j. Then at least one of the small tetrahedra is numbered 1234.

117. Use mathematical induction to prove the following somewhat stronger assertion: If an arbitrary polygon M is divided into triangles in such a way that no two triangles touch along part of a side of one of them, and if an even number of triangles converges at each vertex, then all the vertices may be numbered 1, 2, or 3 in such a way that the vertices on the boundary of M are all numbered 1 or 2 and the small triangles are all numbered 1, 2, 3.

118. First of all, if there are gaps between two neighboring polygons (filled up by other polygons), then attach the gaps to the polygons. This will give a new decomposition of the square in which the boundary between neighboring polygons consists of a single line. Then consider the polygon M_0 containing the center of the square (the polygon of rank one), the polygons neighboring M_0 (polygons of rank two), the polygons neighboring polygons of rank two (polygons of rank three), and so on. Show that if all the small polygons have no more than five neighbors, then there are no polygons of the fifth or higher ranks. This is impossible, since polygons of the first four ranks cannot touch the sides of the square.

119. Show that if the curve has no chord parallel to AB and of length either a or b, then it has no chord parallel to AB of length $a + b$. For the proof, use the fact that the hypothesis is equivalent to the statement that the curve has no point in common with the congruent curve obtained by a parallel displacement a distance a or b to the right.

For the proof of the second half, suppose a lies between $1/n$ and $1/(n + 1)$ (where n is some integer). Construct a continuous curve having no chord parallel to AB and of length lying strictly between $1/n$ and $1/(n + 1)$. (For the construction of an example the reader should first try small values of n.)

120a. Let AB be a side of the convex polygon M of unit area, and C a point of the polygon at maximal distance from AB. Show that M is contained in a parallelogram of area $\leqq 2$, one of whose sides contains AB and two of whose sides are parallel to AC.

b. Consider a parallelogram $APQR$ circumscribed about the triangle ABC of area one, so that B lies on PQ and C on RQ: show that its area is $\geqq 2$.

121a. Inscribe in the given polygon a triangle $A_1A_2A_3$ of maximum possible area, and consider separately the cases where the area of this triangle is $>\frac{1}{2}$ and $\leqq \frac{1}{2}$.

b. Show first that it is impossible to circumscribe a triangle of area less than two about a square of area 1. Then show successively that the result continues to hold for a rectangle as well as the square, and a parallelogram as well as a rectangle. Use the fact that under orthogonal projection all areas are changed in the same ratio.

122a. Suppose the line l does not intersect M. Let A be a vertex of M as near as possible to l, and B a vertex as far as possible from l. Let l_1', l_0, and l_2' be lines parallel to l dividing AB into four equal parts. Let l_1' (nearest to A) intersect M in P and Q, and let l_2' intersect M in R and S. Show that the area of one of the triangles ARS and BPQ is at least $\frac{3}{8}$ the area of M.

b. Find a triangle inscribed in a regular hexagon, one of whose sides is parallel to the side AB of the hexagon, and of maximum area.

123a. Use induction on the number n of progressions. In the proof, it is useful to know that the greatest common divisor d of two integers a

and b can be written in the form $d = pa + qb$, where p and q are integers (not necessarily positive). The second assertion is proved by producing three progressions, each pair of which has a term in common, but such that all three do not have a term in common.

b. Show that if the common differences of two of the given progressions are incommensurable, then they have only one term in common (which therefore must belong to all the other progressions). If the common differences are all commensurable, then the problem can be reduced to one in which all the progressions consist of integers. The result of part a can then be applied.

124a. Direct attempts to construct sequences of 1's and 2's without repetitions will rapidly lead to a contradiction.

b. We introduce the following notation. Let A be some sequence of 1's and 2's. Then $\bar{A}$ is the sequence obtained from A by replacing each 1 by the pair 12 and each 2 by the pair 21. Consider now the following sequences of 1's and 2's:

$$I_1 = 12,$$
$$I_2 = \bar{I}_1 = 12\ \ 21,$$
$$I_3 = \bar{I}_2 = 12\ \ 21\ \ 21\ \ 12,$$
$$I_4 = \bar{I}_3 = 12\ \ 21\ \ 21\ \ 12\ \ 21\ \ 12\ \ 12\ \ 21,$$
$$I_5 = \bar{I}_4 = 12\ \ 21\ \ 21\ \ 12\ \ 21\ \ 12\ \ 12\ \ 21\ \ 21\ \ 12\ \ 12\ \ 21\ \ 12\ \ 21\ \ 21\ \ 12,$$

.
.
.

Show that none of the sequences I_n contains a block of digits occurring three times in a row.

125a. As in the solution to problem 124b, construct sequences $J_0, J_1 = \bar{J}_0, \ldots$ in which $\bar{J}_n$ is obtained from J_n by substituting blocks of digits for the digits of J_n as follows:

$$0 \to 02$$
$$1 \to 0121$$
$$2 \to 0131$$
$$3 \to 03.$$

Thus

$$J_0 = 01,$$
$$J_1 = \bar{J}_0 = 02\ \ 0121,$$
$$J_2 = \bar{J}_1 = 02\ \ 0131\ \ 02\ \ 0121\ \ 0131\ \ 0121,$$
$$J_3 = \bar{J}_2 = 02\ \ 0131\ \ 02\ \ 0121\ \ 03\ \ 0121\ \ 02\ \ 0131\ \ 02\ \ 0121,$$
$$0131\ \ 0121\ \ 02\ \ 0121\ \ 03\ \ 0121\ \ 02\ \ 0121\ \ 0131\ \ 0121,$$

.
.
.

Show that none of the sequences J_n contains a block of digits occurring twice in a row.

b. We introduce the following notation. For any sequence A of digits 1, 2, 3 we denote by $\hat{A}$ the sequence obtained from A by substituting certain blocks of digits for the digits of A as follows:

If in A a 1 stands in an odd-numbered position, replace it by 123, and if in an even position, by 321. Similarly, if 2 stands in an odd position, replace it by 231; otherwise replace it by 132. If 3 stands in an odd position, replace it by 312; if in an even position, by 213. Consider now the following sequences:

$$K_1 = 123,$$
$$K_2 = \hat{K}_1 = 123 \quad 132 \quad 312,$$
$$K_3 = \hat{K}_2 = 123 \quad 132 \quad 312 \quad 321 \quad 312 \quad 132 \quad 312 \quad 321 \quad 231,$$
$$\vdots$$

Prove that none of the sequences $K_1, K_2, K_3, \ldots$ contains a digit or block of digits occurring twice in a row.

126. Define the number $T = T_n$ by the following construction. First write down n 1's in a row. Since we are not allowed to have two identical sequences of n digits in T, the next digit must be a 0. Now continue writing zeros as long as this is possible, that is, until a further zero would introduce two identical sequences of n digits. At this stage write a 1. Now continue writing zeros for as long as possible. Continue in this way, always writing zeros if possible and ones if not, until we are stuck (that is, until we reach a point where whatever we write we obtain two identical n-sequences). At this stage we have completed the construction of T_n.

For clarity we give a number of examples.

$$T_2 = 11001.$$

After the two initial 1's we write two zeros, after which we cannot write any more zeros, and so write a 1. At this stage T_2 stops; we cannot write a zero (or we get two sequences 10), and we cannot write a 1 (or we get two sequences 11). In exactly the same way we can construct the sequences

$$T_3 = 1110001011,$$

$$T_4 = 1111000010011010111,$$

and so on.

It is not difficult to check that every pair (triple, quadruple) occurs once in $T_2(T_3, T_4)$. The proof of the corresponding property of T_n is by induction on the number of 1's at the end of the n-sequence which we are considering.

127. It is convenient to prove the following more general theorem: if there are mn cookies, n each of m different flavors, and if they are put into m boxes so that each box contains n cookies, then it is possible to choose one cookie from each box so that the chosen cookies all have different flavors. For $n = 1$ and $n = 2$ this theorem is almost obvious. Prove it by induction on n.

128. Assume that the theorem is true for $< m$ boys. Now consider separately the following two cases:

(1) Any k boys, where $k < m$, have at least $k + 1$ acquaintances.
(2) There is a set of k boys ($k < m$) who have exactly k acquaintances.

In each case use the induction hypothesis.

129. Let A and B be two nonnegative integers. We write them to base 2 in the form $A = \langle a_n a_{n-1} \cdots a_0 \rangle$, $B = \langle b_n b_{n-1} \cdots b_0 \rangle$. This means that $A = a_n 2^n + a_{n-1} 2^{n-1} + \cdots + a_0$, and $B = b_n 2^n + b_{n-1} 2^{n-1} + \cdots + b_0$, where the digits a_i and b_i are either 0 or 1 (we can suppose that at least one of the "leading digits" a_n, b_n, is equal to 1, but they need not both be 1). We define the *Nim sum* $A \oplus B$ to be the integer $C = \langle c_n c_{n-1} \cdots c_0 \rangle$, where c_i is 0 or 1 according as $a_i + b_i$ is even or odd. For example, if $A = 27$, $B = 13$, then $A = \langle 11011 \rangle$, $B = \langle 01101 \rangle$, so $C = \langle 10110 \rangle = 22$. The Nim sum is obtained by adding A and B in the base 2, but without "carrying".

Show that if the squares of the board are numbered as described in the problem, then the number in the $(A + 1)$st row and the $(B + 1)$st column is $A \oplus B$.

130. Suppose the three piles contain a, b, and c matches, respectively. Write these numbers to base 2 and consider the sum of the last digits, the sum of the next to last digits, etc. If at least one of these sums is odd, then the first player can win. If not, then he loses to correct play by his opponent.

131. Let x be the number of matches in the first pile and y the number in the second pile. Suppose for definiteness that $x \leqq y$. Expand x and y in the F-system as explained on pp. 15–16. Then the first player loses if and only if x ends in an even number of zeros (possibly none), and $y = x0$ (meaning that the expansion of y is obtained from that of x by adding a zero at the end). For example, the first few losing positions are

x	1	3	4	6
y	2	5	7	10

or in the F-system,

x	1	100	101	1001
y	10	1000	1010	10010.

132. Use De Moivre's formula

$$(\cos \alpha + i \sin \alpha)^n = \cos n\alpha + i \sin n\alpha.$$

133. Use the formula of problem 132**b**.

To solve the second part of the problem, use the fact that $\cos \varphi$ assumes its maximum and minimum values when φ is a multiple of π, and the value 0 when $\varphi = \pi/2 + k\pi$, where k is any integer (not necessarily positive).

134. The polynomial $x^2 - \frac{1}{2}$, whose deviation from zero is $\frac{1}{2}$.

135. Consider the polynomial

$$R(x) = P_n(x) - \frac{1}{2^{n-1}} T_n(x),$$

where $P_n(x)$ is any polynomial of degree n with leading coefficient 1 whose deviation from zero on the segment $[-1,+1]$ does not exceed $(\frac{1}{2})^{n-1}$. R is a polynomial of degree $\leq n - 1$. Find points where R must be positive and points where it must be negative, and deduce that the curve $y = R(x)$ cuts the x axis at least n times. This means that the polynomial R of degree less than n has at least n roots. It follows that $R = 0$, that is, $P_n(x) = (\frac{1}{2})^{n-1}T_n(x)$. Examine separately the cases where the deviation from zero of $P_n(x)$ is less than $(\frac{1}{2})^{n-1}$ and equal to $(\frac{1}{2})^{n-1}$.

136. The required polynomials are $2T_n(x/2)$, where n is arbitrary and T_n are the Tchebychev polynomials (see problem 133); the deviation from zero of all of them on the interval $[-2,+2]$ is 2. The result of problem 135 is used in the solution.

137. The polynomial

$$\frac{n!}{2^n}\left\{\frac{(x-1)(x-2)(x-3)\cdots(x-n)}{n!} + \frac{(x-0)(x-2)(x-3)\cdots(x-n)}{1!\,(n-1)!} + \frac{(x-0)(x-1)(x-3)\cdots(x-n)}{2!\,(n-2)!} + \cdots + \frac{(x-0)(x-1)(x-2)\cdots(x-\overline{n-1})}{n!}\right\},$$

whose deviation from zero is $n!/2^n$.

For the proof, use the formula

$$P(x) = (-1)^n P(0)\frac{(x-1)(x-2)(x-3)\cdots(x-n)}{n!}$$
$$+ (-1)^{n-1}P(1)\frac{(x-0)(x-2)(x-3)\cdots(x-n)}{1!\,(n-1)!}$$
$$+ (-1)^{n-2}P(2)\frac{(x-0)(x-1)(x-3)\cdots(x-n)}{2!\,(n-2)!}$$
$$+ \cdots + P(n)\frac{(x-0)(x-1)(x-2)\cdots(x-\overline{n-1})}{n!},$$

where P is an arbitrary polynomial of degree n, and $P(0), P(1), P(2), \ldots, P(n)$ are the values it assumes at the points $0, 1, 2, 3, \ldots, n$, respectively.

138. If there is no point M on a line segment of length l such that $MA_1 \cdot MA_2 \cdots MA_n > 2(l/4)^n$, then $A_1, A_2, \ldots, A_n$ all lie on the segment, at distances $(l/2)\cos\pi/2n$, $(l/2)\cos 3\pi/2n$, $(l/2)\cos 5\pi/2n, \ldots,$ $(l/2)\cos(2n-1)\pi/2n$ from its center.

For the solution, use the geometric representation of complex numbers: if the points $A_1, A_2, \ldots, A_n$ correspond to the complex numbers $\alpha_1, \alpha_2, \ldots, \alpha_n$ and M to the complex number z, then the product $MA_1 \cdot MA_2 \cdots MA_n$ is equal to the absolute value of the polynomial $(z-\alpha_1)(z-\alpha_2)(z-\alpha_3)\cdots(z-\alpha_n)$ of degree n with leading coefficient 1. Now use the result of 136.

139. Use the geometric interpretation of the sine and the tangent as twice the area of certain triangles connected with the unit circle.

140. $(\frac{1}{2})^n \sin\alpha/\sin(\alpha/2^n)$. Multiply the expression by $\sin(\alpha/2^n)$.

141a. $\binom{2m+1}{1}x^m - \binom{2m+1}{3}x^{m-1} + \binom{2m+1}{5}x^{n-2} - \cdots = 0.$

Use the formula of 132**a**, taking $n = 2m + 1$.

b. $x^n - \binom{n}{1}x^{n-1} - \binom{n}{2}x^{n-2} + \binom{n}{3}x^{n-3} + \binom{n}{4}x^{n-4} - \cdots = 0.$

Use the formula for $\cot n\alpha$ which follows from the formula of problem 132**c**.

c, d. The equations

$$\binom{2m}{1}(1-x)^{m-1} - \binom{2m}{3}(1-x)^{m-2}x + \binom{2m}{5}(1-x)^{m-3}x^2 - \cdots = 0$$

and

$$(1-x)^m - \binom{2m}{2}(1-x)^{m-1}x + \binom{2m}{4}(1-x)^{m-2}x^2 - \cdots = 0$$

respectively. Use the formulas of problem 132**a** and **b**, putting $n = 2m$.

142a. Use the result of 141**a**.

b. Follows from the identity of part **a**.

c. Use the result of problem 141**b**.

143. Use the results of problems 141**c** and **d**.

144a. In the identity of problem 140 put $\alpha = \pi/2$ and let $n \to \infty$

b. $\dfrac{3\sqrt{3}}{4\pi}$.

145a. Use the result of problem 139**a**, together with 142**a** and **b**.

b. $\pi^4/90$. Determine the sum of the fourth powers of the cotangents and cosecants of the angles $\pi/(2m + 1), 2\pi/(2m + 1), \ldots, m\pi/(2m + 1)$.

146a. The solution is similar to that of problem 145**a**.

b. $\dfrac{\pi^2}{8}$.

147. Consider the following two expressions, which are suggested by Wallis' formula:

$$\frac{\sin 2\pi/4m}{\sin \pi/4m} \cdot \frac{\sin 2\pi/4m}{\sin 3\pi/4m} \cdot \frac{\sin 4\pi/4m}{\sin 3\pi/4m} \cdot \frac{\sin 4\pi/4m}{\sin 5\pi/4m} \cdots \frac{\sin (2m-2)\pi/4m}{\sin (2m-3)\pi/4m} \times \frac{\sin (2m-2)\pi/4m}{\sin (2m-1)\pi/4m}$$

and

$$\frac{\sin 2\pi/4m}{\sin 3\pi/4m} \cdot \frac{\sin 4\pi/4m}{\sin 3\pi/4m} \cdot \frac{\sin 4\pi/4m}{\sin 5\pi/4m} \cdot \frac{\sin 6\pi/4m}{\sin 5\pi/4m} \cdots \frac{\sin (2m-2)\pi/2m}{\sin (2m-1)\pi/4m} \times \frac{\sin 2m\pi/4m}{\sin (2m-1)\pi/4m}.$$

Then use the result of problem 139**b**.

148. $\dfrac{a^3}{3}$.

149a. 2π.

b. $2 \sin^2 \dfrac{a}{2}$.

150a. $(b^{m+1} - a^{m+1})/(m + 1)$. Divide the side AD of the curvilinear trapezoid $ABCD$ (where $OA = a$, $OD = b$) into segments AM_1, M_1M_2,

$M_2M_3, \ldots, M_{n-1}D$ in such a way that

$$\frac{OM_1}{a} = \frac{OM_2}{OM_1} = \frac{OM_3}{OM_2} = \cdots = \frac{b}{OM_{n-1}}.$$

b. $\dfrac{b^{m+1}}{m+1}$.

151. If $OA_1 = a_1$, $OD_1 = b_1$, $OA_2 = a_2$, $OD_2 = b_2$, divide the segments A_1D_1 and A_2D_2 into n equal parts and replace the curvilinear trapezoid by a stepped figure.

152. Use the result of problem 151.

153. Use the fact that $F(z)$ is a continuous and increasing function of its argument z.

154. Show first that for any α, $F(z^\alpha) = \alpha F(z)$.

155. $3 \ln 2 - 2$ (use problem 154).

156. $2 \ln 2 - 1$. The outcome of the experiment can be described by the point (x,y), where x is the smaller of the two segments obtained by breaking the rod the first time, and y is the fraction then broken off from the bigger segment. (Thus $0 \leqq y \leqq 1$.) The possible outcomes (x,y) form a rectangle R, and the problem is to find the area of the subset of R formed by the favorable outcomes.

157. Apply the result of problem 149**b**.

158a. Use the result of problem 154.

b. $1/\ln a$. Use the result of part **a**.

c. $\ln a$. Use the result of problem 154.

159a. $(a^b - 1)/(\ln a)$. The proof requires the result of problem 158**c**.

b. $(b \ln b - b + 1)/(\ln a)$. This result may be deduced from the formula of part a or established independently by the method used for the solution of 150**a**. (See the hint to that problem.) If the second method is used, the result of problem 158**c** will be needed.

160. $\frac{1}{2} \ln a (\log_a b)^2$. The solution to this problem is analogous to the second solution of problem 159**b**.

161. Use the result of problem 150**b**.

162a. It is sufficient to show that if $n > m$, then $n \ln (1 + 1/n) > m \ln (1 + 1/m)$. Use the result of problem 154.

b. Show that $(n + 1) \ln (1 + 1/n) > (n + 2) \ln (1 + 1/(n + 1))$; use the result of problem 154.

163. Consider separately the cases where z is positive and negative; use the result of problem 154.

164. Use the result of problem 163.

165. Evaluate the area of the curvilinear triangle bounded by the x axis, the curve $y = \ln x$, and the line $x = n$ using two different methods. In one case start by considering the area of a trapezoid inscribed in the curve, in the other, circumscribed. Use the result of problem 159**b**.

166a. The solution is based on that of 165.
b. Use Wallis' formula (problem 147).

167. Use the result of problem 154.

168. Use the result of problem 160.

169. Use the result of problem 150**a**.

170. Estimate by two different methods the value of the quantity $\binom{2n}{n} = \frac{(2n)!}{(n!)^2}$. Specifically, show that

$$2^n \leqq \binom{2n}{n} < 2^{2n}$$

and

$$n^{\pi(2n)-\pi(n)} < \binom{2n}{n} \leqq (2n)^{\pi(2n)}.$$

171. Let $p_1, p_2, \ldots, p_r$ be all the primes not exceeding the integer N; suppose the prime decomposition of $N!$ is of the form $N! = p_1^{\alpha_1} p_2^{\alpha_2} \cdots p_r^{\alpha_r}$. Then, as may easily be shown,

$$\alpha_i = \left[\frac{N}{p_i}\right] + \left[\frac{N}{p_i^2}\right] + \left[\frac{N}{p_i^3}\right] + \cdots,$$

where $[x]$ is the integer part of x. Using this expression, we may obtain an estimate for $\log N!$ in which the sum $(\log 2)/2 + (\log 3)/3 + (\log 5)/5 + \cdots + (\log p)/p$ required by the problem appears. It will also be necessary to use Tchebychev's theorem (problem 170).

The second estimate for $\log N!$ may be obtained from the result of problem 165. Finally, compare the two estimates obtained for $\log N!$.

172a. Express the numbers $b_1, b_2, \ldots, b_n$ in terms of the numbers $B_1, B_2, \ldots, B_n$ and substitute for them in the sum S.

b. (1) $\dfrac{nq^n}{q-1} - \dfrac{q^n - 1}{(q-1)^2}$;

(2) $\dfrac{n^2q^n}{q-1} - \dfrac{(2n-1)q^n + 1}{(q-1)^2} + \dfrac{2q^n - 2}{(q-1)^3}$.

173. Apply the formula of problem 172a to the sum $\frac{1}{2} + \frac{1}{3} + \frac{1}{5} + \frac{1}{7} + \cdots + 1/p$ and use Mertens' first theorem (problem 171). It will also prove necessary to use certain estimates similar to those of problem 167.

174. Clearly,

$$\ln\left[\left(1-\frac{1}{2}\right)\left(1-\frac{1}{3}\right)\left(1-\frac{1}{5}\right)\left(1-\frac{1}{7}\right)\left(1-\frac{1}{11}\right)\cdots\left(1-\frac{1}{p}\right)\right]$$

$$= \ln\left(1-\frac{1}{2}\right) + \ln\left(1-\frac{1}{3}\right) + \ln\left(1-\frac{1}{5}\right) + \ln\left(1-\frac{1}{7}\right)$$

$$+ \ln\left(1-\frac{1}{11}\right) + \cdots + \ln\left(1-\frac{1}{p}\right).$$

To obtain an estimate for the right-hand side, use the geometric definition of the natural logarithm (see problems 151–154). Finally, use Mertens' second theorem (problem 173).

BIBLIOGRAPHY

1 Abel'son, I. B., Rozhdenie logarifmov (transl: The Birth of Logarithms), Gostekhizdat, Moscow and Leningrad, 1948.

2 Alexsandrov, P. S., Elementary Concepts of Topology, Dover, New York, 1961.

3 Beckenbach, E. and R. Bellman, An Introduction to Inequalities, Random House, New York, 1961.

4 Coxeter, H. S. M., The Real Projective Plane, Cambridge, London, 1960.

5 Coxeter, H. S. M., Scripta Math. 19, 135 (1953).

6 Coxeter, H. S. M., Am. Math. Monthly 55, 26 (1948).

7 Dynkin, E. B. and V. A. Uspenskii, Mathematical Conversations, Vol. I, Heath, Boston, 1962.

8 Halmos, P. and H. Vaughan, Am. J. Math. 72, 214 (1950).

9 Hardy, G. H., and E. M. Wright, An Introduction to the Theory of Numbers, Clarendon Press, Oxford, 1954, chap. XIII.

10 Hilbert, D. and S. Cohn-Vossen, Geometry and the Imagination, Chelsea, New York, 1952.

11 Hobson, E. W., Squaring the Circle, a History of the Problem, Cambridge University Press, London, 1913.

12 Ingham, A. E., The Distribution of Prime Numbers, Stechert-Hafner, New York, 1964.

13 Johnson, E. L., Am. Math. Monthly 73, 52 (1966).

14 Kelly, L. M. and W. O. J. Moser, Can. J. Math. 2, 210 (1958).

15 Knopp, K., Theory of Functions, Dover, New York, vol. 1, p. 17.

16 Konig, D., Theorie der Endlichen und Unendlichen Graphen, Chelsea, New York, 1950, pp. 171–174.

17 Krechmar, V. A., Zadachnik po algebre (transl: Problem Book in Algebra), Gostekhizdat, Moscow and Leningrad, 1950, chap. X.

18 Le Veque, W. J., Topics in Number Theory, Addison-Wesley, Reading, 1956.

19 Lyusternik, L. A., Convex Figures and Polyhedra, Dover, New York, 1963.

20 Markushevich, A. I., Areas and Logarithms, Heath, Boston, 1963.

21 Shklyarskii, D. O., N. N. Chentsov, and I. M. Yaglom, Izbrannye zadachi i teoremy elementarnoi (transl: Selected Problems and Theorems in Elementary Mathematics), Gostekhizdat, Moscow, 1954, chap. I, cycle 6.

22 Valentine, F., Convex Sets, McGraw-Hill, New York, 1964.

23 Vinogradov, I. M., An Introduction to the Theory of Numbers, Pergamon Press, London, 1955.

24 Yaglom, I. M. and G. Boltyanskii, Convex Figures, Holt, Rinehart and Winston, New York, 1961.

25 Wythoff, I. I., Nieuw Arch. v. Wisk. 7, 199 (1907).
26 Hopf, H., Comm. Math. Helv. 9, 303 (1937).
27 Levi, F. W., J. Indian. Math. Soc. 15, 65 (1951).

A CATALOG OF SELECTED

DOVER BOOKS

IN SCIENCE AND MATHEMATICS

Astronomy

CHARIOTS FOR APOLLO: The NASA History of Manned Lunar Spacecraft to 1969, Courtney G. Brooks, James M. Grimwood, and Loyd S. Swenson, Jr. This illustrated history by a trio of experts is the definitive reference on the Apollo spacecraft and lunar modules. It traces the vehicles' design, development, and operation in space. More than 100 photographs and illustrations. 576pp. 6 3/4 x 9 1/4. 0-486-46756-2

EXPLORING THE MOON THROUGH BINOCULARS AND SMALL TELESCOPES, Ernest H. Cherrington, Jr. Informative, profusely illustrated guide to locating and identifying craters, rills, seas, mountains, other lunar features. Newly revised and updated with special section of new photos. Over 100 photos and diagrams. 240pp. 8 1/4 x 11. 0-486-24491-1

WHERE NO MAN HAS GONE BEFORE: A History of NASA's Apollo Lunar Expeditions, William David Compton. Introduction by Paul Dickson. This official NASA history traces behind-the-scenes conflicts and cooperation between scientists and engineers. The first half concerns preparations for the Moon landings, and the second half documents the flights that followed Apollo 11. 1989 edition. 432pp. 7 x 10. 0-486-47888-2

APOLLO EXPEDITIONS TO THE MOON: The NASA History, Edited by Edgar M. Cortright. Official NASA publication marks the 40th anniversary of the first lunar landing and features essays by project participants recalling engineering and administrative challenges. Accessible, jargon-free accounts, highlighted by numerous illustrations. 336pp. 8 3/8 x 10 7/8. 0-486-47175-6

ON MARS: Exploration of the Red Planet, 1958-1978--The NASA History, Edward Clinton Ezell and Linda Neuman Ezell. NASA's official history chronicles the start of our explorations of our planetary neighbor. It recounts cooperation among government, industry, and academia, and it features dozens of photos from Viking cameras. 560pp. 6 3/4 x 9 1/4. 0-486-46757-0

ARISTARCHUS OF SAMOS: The Ancient Copernicus, Sir Thomas Heath. Heath's history of astronomy ranges from Homer and Hesiod to Aristarchus and includes quotes from numerous thinkers, compilers, and scholasticists from Thales and Anaximander through Pythagoras, Plato, Aristotle, and Heraclides. 34 figures. 448pp. 5 3/8 x 8 1/2. 0-486-43886-4

AN INTRODUCTION TO CELESTIAL MECHANICS, Forest Ray Moulton. Classic text still unsurpassed in presentation of fundamental principles. Covers rectilinear motion, central forces, problems of two and three bodies, much more. Includes over 200 problems, some with answers. 437pp. 5 3/8 x 8 1/2. 0-486-64687-4

BEYOND THE ATMOSPHERE: Early Years of Space Science, Homer E. Newell. This exciting survey is the work of a top NASA administrator who chronicles technological advances, the relationship of space science to general science, and the space program's social, political, and economic contexts. 528pp. 6 3/4 x 9 1/4. 0-486-47464-X

STAR LORE: Myths, Legends, and Facts, William Tyler Olcott. Captivating retellings of the origins and histories of ancient star groups include Pegasus, Ursa Major, Pleiades, signs of the zodiac, and other constellations. "Classic." – *Sky & Telescope.* 58 illustrations. 544pp. 5 3/8 x 8 1/2. 0-486-43581-4

A COMPLETE MANUAL OF AMATEUR ASTRONOMY: Tools and Techniques for Astronomical Observations, P. Clay Sherrod with Thomas L. Koed. Concise, highly readable book discusses the selection, set-up, and maintenance of a telescope; amateur studies of the sun; lunar topography and occultations; and more. 124 figures. 26 halftones. 37 tables. 335pp. 6 1/2 x 9 1/4. 0-486-42820-6

Chemistry

MOLECULAR COLLISION THEORY, M. S. Child. This high-level monograph offers an analytical treatment of classical scattering by a central force, quantum scattering by a central force, elastic scattering phase shifts, and semi-classical elastic scattering. 1974 edition. 310pp. 5 3/8 x 8 1/2. 0-486-69437-2

HANDBOOK OF COMPUTATIONAL QUANTUM CHEMISTRY, David B. Cook. This comprehensive text provides upper-level undergraduates and graduate students with an accessible introduction to the implementation of quantum ideas in molecular modeling, exploring practical applications alongside theoretical explanations. 1998 edition. 832pp. 5 3/8 x 8 1/2. 0-486-44307-8

RADIOACTIVE SUBSTANCES, Marie Curie. The celebrated scientist's thesis, which directly preceded her 1903 Nobel Prize, discusses establishing atomic character of radioactivity; extraction from pitchblende of polonium and radium; isolation of pure radium chloride; more. 96pp. 5 3/8 x 8 1/2. 0-486-42550-9

CHEMICAL MAGIC, Leonard A. Ford. Classic guide provides intriguing entertainment while elucidating sound scientific principles, with more than 100 unusual stunts: cold fire, dust explosions, a nylon rope trick, a disappearing beaker, much more. 128pp. 5 3/8 x 8 1/2. 0-486-67628-5

ALCHEMY, E. J. Holmyard. Classic study by noted authority covers 2,000 years of alchemical history: religious, mystical overtones; apparatus; signs, symbols, and secret terms; advent of scientific method, much more. Illustrated. 320pp. 5 3/8 x 8 1/2. 0-486-26298-7

CHEMICAL KINETICS AND REACTION DYNAMICS, Paul L. Houston. This text teaches the principles underlying modern chemical kinetics in a clear, direct fashion, using several examples to enhance basic understanding. Solutions to selected problems. 2001 edition. 352pp. 8 3/8 x 11. 0-486-45334-0

PROBLEMS AND SOLUTIONS IN QUANTUM CHEMISTRY AND PHYSICS, Charles S. Johnson and Lee G. Pedersen. Unusually varied problems, with detailed solutions, cover of quantum mechanics, wave mechanics, angular momentum, molecular spectroscopy, scattering theory, more. 280 problems, plus 139 supplementary exercises. 430pp. 6 1/2 x 9 1/4. 0-486-65236-X

ELEMENTS OF CHEMISTRY, Antoine Lavoisier. Monumental classic by the founder of modern chemistry features first explicit statement of law of conservation of matter in chemical change, and more. Facsimile reprint of original (1790) Kerr translation. 539pp. 5 3/8 x 8 1/2. 0-486-64624-6

MAGNETISM AND TRANSITION METAL COMPLEXES, F. E. Mabbs and D. J. Machin. A detailed view of the calculation methods involved in the magnetic properties of transition metal complexes, this volume offers sufficient background for original work in the field. 1973 edition. 240pp. 5 3/8 x 8 1/2. 0-486-46284-6

GENERAL CHEMISTRY, Linus Pauling. Revised third edition of classic first-year text by Nobel laureate. Atomic and molecular structure, quantum mechanics, statistical mechanics, thermodynamics correlated with descriptive chemistry. Problems. 992pp. 5 3/8 x 8 1/2. 0-486-65622-5

ELECTROLYTE SOLUTIONS: Second Revised Edition, R. A. Robinson and R. H. Stokes. Classic text deals primarily with measurement, interpretation of conductance, chemical potential, and diffusion in electrolyte solutions. Detailed theoretical interpretations, plus extensive tables of thermodynamic and transport properties. 1970 edition. 590pp. 5 3/8 x 8 1/2. 0-486-42225-9

Engineering

FUNDAMENTALS OF ASTRODYNAMICS, Roger R. Bate, Donald D. Mueller, and Jerry E. White. Teaching text developed by U.S. Air Force Academy develops the basic two-body and n-body equations of motion; orbit determination; classical orbital elements, coordinate transformations; differential correction; more. 1971 edition. 455pp. 5 3/8 x 8 1/2. 0-486-60061-0

INTRODUCTION TO CONTINUUM MECHANICS FOR ENGINEERS: Revised Edition, Ray M. Bowen. This self-contained text introduces classical continuum models within a modern framework. Its numerous exercises illustrate the governing principles, linearizations, and other approximations that constitute classical continuum models. 2007 edition. 320pp. 6 1/8 x 9 1/4. 0-486-47460-7

ENGINEERING MECHANICS FOR STRUCTURES, Louis L. Bucciarelli. This text explores the mechanics of solids and statics as well as the strength of materials and elasticity theory. Its many design exercises encourage creative initiative and systems thinking. 2009 edition. 320pp. 6 1/8 x 9 1/4. 0-486-46855-0

FEEDBACK CONTROL THEORY, John C. Doyle, Bruce A. Francis and Allen R. Tannenbaum. This excellent introduction to feedback control system design offers a theoretical approach that captures the essential issues and can be applied to a wide range of practical problems. 1992 edition. 224pp. 6 1/2 x 9 1/4. 0-486-46933-6

THE FORCES OF MATTER, Michael Faraday. These lectures by a famous inventor offer an easy-to-understand introduction to the interactions of the universe's physical forces. Six essays explore gravitation, cohesion, chemical affinity, heat, magnetism, and electricity. 1993 edition. 96pp. 5 3/8 x 8 1/2. 0-486-47482-8

DYNAMICS, Lawrence E. Goodman and William H. Warner. Beginning engineering text introduces calculus of vectors, particle motion, dynamics of particle systems and plane rigid bodies, technical applications in plane motions, and more. Exercises and answers in every chapter. 619pp. 5 3/8 x 8 1/2. 0-486-42006-X

ADAPTIVE FILTERING PREDICTION AND CONTROL, Graham C. Goodwin and Kwai Sang Sin. This unified survey focuses on linear discrete-time systems and explores natural extensions to nonlinear systems. It emphasizes discrete-time systems, summarizing theoretical and practical aspects of a large class of adaptive algorithms. 1984 edition. 560pp. 6 1/2 x 9 1/4. 0-486-46932-8

INDUCTANCE CALCULATIONS, Frederick W. Grover. This authoritative reference enables the design of virtually every type of inductor. It features a single simple formula for each type of inductor, together with tables containing essential numerical factors. 1946 edition. 304pp. 5 3/8 x 8 1/2. 0-486-47440-2

THERMODYNAMICS: Foundations and Applications, Elias P. Gyftopoulos and Gian Paolo Beretta. Designed by two MIT professors, this authoritative text discusses basic concepts and applications in detail, emphasizing generality, definitions, and logical consistency. More than 300 solved problems cover realistic energy systems and processes. 800pp. 6 1/8 x 9 1/4. 0-486-43932-1

THE FINITE ELEMENT METHOD: Linear Static and Dynamic Finite Element Analysis, Thomas J. R. Hughes. Text for students without in-depth mathematical training, this text includes a comprehensive presentation and analysis of algorithms of time-dependent phenomena plus beam, plate, and shell theories. Solution guide available upon request. 672pp. 6 1/2 x 9 1/4. 0-486-41181-8

HELICOPTER THEORY, Wayne Johnson. Monumental engineering text covers vertical flight, forward flight, performance, mathematics of rotating systems, rotary wing dynamics and aerodynamics, aeroelasticity, stability and control, stall, noise, and more. 189 illustrations. 1980 edition. 1089pp. 5 5/8 x 8 1/4. 0-486-68230-7

MATHEMATICAL HANDBOOK FOR SCIENTISTS AND ENGINEERS: Definitions, Theorems, and Formulas for Reference and Review, Granino A. Korn and Theresa M. Korn. Convenient access to information from every area of mathematics: Fourier transforms, Z transforms, linear and nonlinear programming, calculus of variations, random-process theory, special functions, combinatorial analysis, game theory, much more. 1152pp. 5 3/8 x 8 1/2. 0-486-41147-8

A HEAT TRANSFER TEXTBOOK: Fourth Edition, John H. Lienhard V and John H. Lienhard IV. This introduction to heat and mass transfer for engineering students features worked examples and end-of-chapter exercises. Worked examples and end-of-chapter exercises appear throughout the book, along with well-drawn, illuminating figures. 768pp. 7 x 9 1/4. 0-486-47931-5

BASIC ELECTRICITY, U.S. Bureau of Naval Personnel. Originally a training course; best nontechnical coverage. Topics include batteries, circuits, conductors, AC and DC, inductance and capacitance, generators, motors, transformers, amplifiers, etc. Many questions with answers. 349 illustrations. 1969 edition. 448pp. 6 1/2 x 9 1/4. 0-486-20973-3

BASIC ELECTRONICS, U.S. Bureau of Naval Personnel. Clear, well-illustrated introduction to electronic equipment covers numerous essential topics: electron tubes, semiconductors, electronic power supplies, tuned circuits, amplifiers, receivers, ranging and navigation systems, computers, antennas, more. 560 illustrations. 567pp. 6 1/2 x 9 1/4. 0-486-21076-6

BASIC WING AND AIRFOIL THEORY, Alan Pope. This self-contained treatment by a pioneer in the study of wind effects covers flow functions, airfoil construction and pressure distribution, finite and monoplane wings, and many other subjects. 1951 edition. 320pp. 5 3/8 x 8 1/2. 0-486-47188-8

SYNTHETIC FUELS, Ronald F. Probstein and R. Edwin Hicks. This unified presentation examines the methods and processes for converting coal, oil, shale, tar sands, and various forms of biomass into liquid, gaseous, and clean solid fuels. 1982 edition. 512pp. 6 1/8 x 9 1/4. 0-486-44977-7

THEORY OF ELASTIC STABILITY, Stephen P. Timoshenko and James M. Gere. Written by world-renowned authorities on mechanics, this classic ranges from theoretical explanations of 2- and 3-D stress and strain to practical applications such as torsion, bending, and thermal stress. 1961 edition. 560pp. 5 3/8 x 8 1/2. 0-486-47207-8

PRINCIPLES OF DIGITAL COMMUNICATION AND CODING, Andrew J. Viterbi and Jim K. Omura. This classic by two digital communications experts is geared toward students of communications theory and to designers of channels, links, terminals, modems, or networks used to transmit and receive digital messages. 1979 edition. 576pp. 6 1/8 x 9 1/4. 0-486-46901-8

LINEAR SYSTEM THEORY: The State Space Approach, Lotfi A. Zadeh and Charles A. Desoer. Written by two pioneers in the field, this exploration of the state space approach focuses on problems of stability and control, plus connections between this approach and classical techniques. 1963 edition. 656pp. 6 1/8 x 9 1/4. 0-486-46663-9

Mathematics–Bestsellers

HANDBOOK OF MATHEMATICAL FUNCTIONS: with Formulas, Graphs, and Mathematical Tables, Edited by Milton Abramowitz and Irene A. Stegun. A classic resource for working with special functions, standard trig, and exponential logarithmic definitions and extensions, it features 29 sets of tables, some to as high as 20 places. 1046pp. 8 x 10 1/2. 0-486-61272-4

ABSTRACT AND CONCRETE CATEGORIES: The Joy of Cats, Jiri Adamek, Horst Herrlich, and George E. Strecker. This up-to-date introductory treatment employs category theory to explore the theory of structures. Its unique approach stresses concrete categories and presents a systematic view of factorization structures. Numerous examples. 1990 edition, updated 2004. 528pp. 6 1/8 x 9 1/4. 0-486-46934-4

MATHEMATICS: Its Content, Methods and Meaning, A. D. Aleksandrov, A. N. Kolmogorov, and M. A. Lavrent'ev. Major survey offers comprehensive, coherent discussions of analytic geometry, algebra, differential equations, calculus of variations, functions of a complex variable, prime numbers, linear and non-Euclidean geometry, topology, functional analysis, more. 1963 edition. 1120pp. 5 3/8 x 8 1/2. 0-486-40916-3

INTRODUCTION TO VECTORS AND TENSORS: Second Edition--Two Volumes Bound as One, Ray M. Bowen and C.-C. Wang. Convenient single-volume compilation of two texts offers both introduction and in-depth survey. Geared toward engineering and science students rather than mathematicians, it focuses on physics and engineering applications. 1976 edition. 560pp. 6 1/2 x 9 1/4. 0-486-46914-X

AN INTRODUCTION TO ORTHOGONAL POLYNOMIALS, Theodore S. Chihara. Concise introduction covers general elementary theory, including the representation theorem and distribution functions, continued fractions and chain sequences, the recurrence formula, special functions, and some specific systems. 1978 edition. 272pp. 5 3/8 x 8 1/2. 0-486-47929-3

ADVANCED MATHEMATICS FOR ENGINEERS AND SCIENTISTS, Paul DuChateau. This primary text and supplemental reference focuses on linear algebra, calculus, and ordinary differential equations. Additional topics include partial differential equations and approximation methods. Includes solved problems. 1992 edition. 400pp. 7 1/2 x 9 1/4. 0-486-47930-7

PARTIAL DIFFERENTIAL EQUATIONS FOR SCIENTISTS AND ENGINEERS, Stanley J. Farlow. Practical text shows how to formulate and solve partial differential equations. Coverage of diffusion-type problems, hyperbolic-type problems, elliptic-type problems, numerical and approximate methods. Solution guide available upon request. 1982 edition. 414pp. 6 1/8 x 9 1/4. 0-486-67620-X

VARIATIONAL PRINCIPLES AND FREE-BOUNDARY PROBLEMS, Avner Friedman. Advanced graduate-level text examines variational methods in partial differential equations and illustrates their applications to free-boundary problems. Features detailed statements of standard theory of elliptic and parabolic operators. 1982 edition. 720pp. 6 1/8 x 9 1/4. 0-486-47853-X

LINEAR ANALYSIS AND REPRESENTATION THEORY, Steven A. Gaal. Unified treatment covers topics from the theory of operators and operator algebras on Hilbert spaces; integration and representation theory for topological groups; and the theory of Lie algebras, Lie groups, and transform groups. 1973 edition. 704pp. 6 1/8 x 9 1/4. 0-486-47851-3

Mathematics–Logic and Problem Solving

PERPLEXING PUZZLES AND TANTALIZING TEASERS, Martin Gardner. Ninety-three riddles, mazes, illusions, tricky questions, word and picture puzzles, and other challenges offer hours of entertainment for youngsters. Filled with rib-tickling drawings. Solutions. 224pp. 5 3/8 x 8 1/2. 0-486-25637-5

MY BEST MATHEMATICAL AND LOGIC PUZZLES, Martin Gardner. The noted expert selects 70 of his favorite "short" puzzles. Includes The Returning Explorer, The Mutilated Chessboard, Scrambled Box Tops, and dozens more. Complete solutions included. 96pp. 5 3/8 x 8 1/2. 0-486-28152-3

THE LADY OR THE TIGER?: and Other Logic Puzzles, Raymond M. Smullyan. Created by a renowned puzzle master, these whimsically themed challenges involve paradoxes about probability, time, and change; metapuzzles; and self-referentiality. Nineteen chapters advance in difficulty from relatively simple to highly complex. 1982 edition. 240pp. 5 3/8 x 8 1/2. 0-486-47027-X

SATAN, CANTOR AND INFINITY: Mind-Boggling Puzzles, Raymond M. Smullyan. A renowned mathematician tells stories of knights and knaves in an entertaining look at the logical precepts behind infinity, probability, time, and change. Requires a strong background in mathematics. Complete solutions. 288pp. 5 3/8 x 8 1/2. 0-486-47036-9

THE RED BOOK OF MATHEMATICAL PROBLEMS, Kenneth S. Williams and Kenneth Hardy. Handy compilation of 100 practice problems, hints and solutions indispensable for students preparing for the William Lowell Putnam and other mathematical competitions. Preface to the First Edition. Sources. 1988 edition. 192pp. 5 3/8 x 8 1/2. 0-486-69415-1

KING ARTHUR IN SEARCH OF HIS DOG AND OTHER CURIOUS PUZZLES, Raymond M. Smullyan. This fanciful, original collection for readers of all ages features arithmetic puzzles, logic problems related to crime detection, and logic and arithmetic puzzles involving King Arthur and his Dogs of the Round Table. 160pp. 5 3/8 x 8 1/2. 0-486-47435-6

UNDECIDABLE THEORIES: Studies in Logic and the Foundation of Mathematics, Alfred Tarski in collaboration with Andrzej Mostowski and Raphael M. Robinson. This well-known book by the famed logician consists of three treatises: "A General Method in Proofs of Undecidability," "Undecidability and Essential Undecidability in Mathematics," and "Undecidability of the Elementary Theory of Groups." 1953 edition. 112pp. 5 3/8 x 8 1/2. 0-486-47703-7

LOGIC FOR MATHEMATICIANS, J. Barkley Rosser. Examination of essential topics and theorems assumes no background in logic. "Undoubtedly a major addition to the literature of mathematical logic." – *Bulletin of the American Mathematical Society.* 1978 edition. 592pp. 6 1/8 x 9 1/4. 0-486-46898-4

INTRODUCTION TO PROOF IN ABSTRACT MATHEMATICS, Andrew Wohlgemuth. This undergraduate text teaches students what constitutes an acceptable proof, and it develops their ability to do proofs of routine problems as well as those requiring creative insights. 1990 edition. 384pp. 6 1/2 x 9 1/4. 0-486-47854-8

FIRST COURSE IN MATHEMATICAL LOGIC, Patrick Suppes and Shirley Hill. Rigorous introduction is simple enough in presentation and context for wide range of students. Symbolizing sentences; logical inference; truth and validity; truth tables; terms, predicates, universal quantifiers; universal specification and laws of identity; more. 288pp. 5 3/8 x 8 1/2. 0-486-42259-3

Mathematics–Algebra and Calculus

VECTOR CALCULUS, Peter Baxandall and Hans Liebeck. This introductory text offers a rigorous, comprehensive treatment. Classical theorems of vector calculus are amply illustrated with figures, worked examples, physical applications, and exercises with hints and answers. 1986 edition. 560pp. 5 3/8 x 8 1/2. 0-486-46620-5

ADVANCED CALCULUS: An Introduction to Classical Analysis, Louis Brand. A course in analysis that focuses on the functions of a real variable, this text introduces the basic concepts in their simplest setting and illustrates its teachings with numerous examples, theorems, and proofs. 1955 edition. 592pp. 5 3/8 x 8 1/2. 0-486-44548-8

ADVANCED CALCULUS, Avner Friedman. Intended for students who have already completed a one-year course in elementary calculus, this two-part treatment advances from functions of one variable to those of several variables. Solutions. 1971 edition. 432pp. 5 3/8 x 8 1/2. 0-486-45795-8

METHODS OF MATHEMATICS APPLIED TO CALCULUS, PROBABILITY, AND STATISTICS, Richard W. Hamming. This 4-part treatment begins with algebra and analytic geometry and proceeds to an exploration of the calculus of algebraic functions and transcendental functions and applications. 1985 edition. Includes 310 figures and 18 tables. 880pp. 6 1/2 x 9 1/4. 0-486-43945-3

BASIC ALGEBRA I: Second Edition, Nathan Jacobson. A classic text and standard reference for a generation, this volume covers all undergraduate algebra topics, including groups, rings, modules, Galois theory, polynomials, linear algebra, and associative algebra. 1985 edition. 528pp. 6 1/8 x 9 1/4. 0-486-47189-6

BASIC ALGEBRA II: Second Edition, Nathan Jacobson. This classic text and standard reference comprises all subjects of a first-year graduate-level course, including in-depth coverage of groups and polynomials and extensive use of categories and functors. 1989 edition. 704pp. 6 1/8 x 9 1/4. 0-486-47187-X

CALCULUS: An Intuitive and Physical Approach (Second Edition), Morris Kline. Application-oriented introduction relates the subject as closely as possible to science with explorations of the derivative; differentiation and integration of the powers of x; theorems on differentiation, antidifferentiation; the chain rule; trigonometric functions; more. Examples. 1967 edition. 960pp. 6 1/2 x 9 1/4. 0-486-40453-6

ABSTRACT ALGEBRA AND SOLUTION BY RADICALS, John E. Maxfield and Margaret W. Maxfield. Accessible advanced undergraduate-level text starts with groups, rings, fields, and polynomials and advances to Galois theory, radicals and roots of unity, and solution by radicals. Numerous examples, illustrations, exercises, appendixes. 1971 edition. 224pp. 6 1/8 x 9 1/4. 0-486-47723-1

AN INTRODUCTION TO THE THEORY OF LINEAR SPACES, Georgi E. Shilov. Translated by Richard A. Silverman. Introductory treatment offers a clear exposition of algebra, geometry, and analysis as parts of an integrated whole rather than separate subjects. Numerous examples illustrate many different fields, and problems include hints or answers. 1961 edition. 320pp. 5 3/8 x 8 1/2. 0-486-63070-6

LINEAR ALGEBRA, Georgi E. Shilov. Covers determinants, linear spaces, systems of linear equations, linear functions of a vector argument, coordinate transformations, the canonical form of the matrix of a linear operator, bilinear and quadratic forms, and more. 387pp. 5 3/8 x 8 1/2. 0-486-63518-X

Mathematics–Probability and Statistics

BASIC PROBABILITY THEORY, Robert B. Ash. This text emphasizes the probabilistic way of thinking, rather than measure-theoretic concepts. Geared toward advanced undergraduates and graduate students, it features solutions to some of the problems. 1970 edition. 352pp. 5 3/8 x 8 1/2. 0-486-46628-0

PRINCIPLES OF STATISTICS, M. G. Bulmer. Concise description of classical statistics, from basic dice probabilities to modern regression analysis. Equal stress on theory and applications. Moderate difficulty; only basic calculus required. Includes problems with answers. 252pp. 5 5/8 x 8 1/4. 0-486-63760-3

OUTLINE OF BASIC STATISTICS: Dictionary and Formulas, John E. Freund and Frank J. Williams. Handy guide includes a 70-page outline of essential statistical formulas covering grouped and ungrouped data, finite populations, probability, and more, plus over 1,000 clear, concise definitions of statistical terms. 1966 edition. 208pp. 5 3/8 x 8 1/2. 0-486-47769-X

GOOD THINKING: The Foundations of Probability and Its Applications, Irving J. Good. This in-depth treatment of probability theory by a famous British statistician explores Keynesian principles and surveys such topics as Bayesian rationality, corroboration, hypothesis testing, and mathematical tools for induction and simplicity. 1983 edition. 352pp. 5 3/8 x 8 1/2. 0-486-47438-0

INTRODUCTION TO PROBABILITY THEORY WITH CONTEMPORARY APPLICATIONS, Lester L. Helms. Extensive discussions and clear examples, written in plain language, expose students to the rules and methods of probability. Exercises foster problem-solving skills, and all problems feature step-by-step solutions. 1997 edition. 368pp. 6 1/2 x 9 1/4. 0-486-47418-6

CHANCE, LUCK, AND STATISTICS, Horace C. Levinson. In simple, non-technical language, this volume explores the fundamentals governing chance and applies them to sports, government, and business. "Clear and lively ... remarkably accurate." – *Scientific Monthly*. 384pp. 5 3/8 x 8 1/2. 0-486-41997-5

FIFTY CHALLENGING PROBLEMS IN PROBABILITY WITH SOLUTIONS, Frederick Mosteller. Remarkable puzzlers, graded in difficulty, illustrate elementary and advanced aspects of probability. These problems were selected for originality, general interest, or because they demonstrate valuable techniques. Also includes detailed solutions. 88pp. 5 3/8 x 8 1/2. 0-486-65355-2

EXPERIMENTAL STATISTICS, Mary Gibbons Natrella. A handbook for those seeking engineering information and quantitative data for designing, developing, constructing, and testing equipment. Covers the planning of experiments, the analyzing of extreme-value data; and more. 1966 edition. Index. Includes 52 figures and 76 tables. 560pp. 8 3/8 x 11. 0-486-43937-2

STOCHASTIC MODELING: Analysis and Simulation, Barry L. Nelson. Coherent introduction to techniques also offers a guide to the mathematical, numerical, and simulation tools of systems analysis. Includes formulation of models, analysis, and interpretation of results. 1995 edition. 336pp. 6 1/8 x 9 1/4. 0-486-47770-3

INTRODUCTION TO BIOSTATISTICS: Second Edition, Robert R. Sokal and F. James Rohlf. Suitable for undergraduates with a minimal background in mathematics, this introduction ranges from descriptive statistics to fundamental distributions and the testing of hypotheses. Includes numerous worked-out problems and examples. 1987 edition. 384pp. 6 1/8 x 9 1/4. 0-486-46961-1

Mathematics–Geometry and Topology

Mathematics-History

THE WORKS OF ARCHIMEDES, Archimedes. Translated by Sir Thomas Heath. Complete works of ancient geometer feature such topics as the famous problems of the ratio of the areas of a cylinder and an inscribed sphere; the properties of conoids, spheroids, and spirals; more. 326pp. 5 3/8 x 8 1/2. 0-486-42084-1

THE HISTORICAL ROOTS OF ELEMENTARY MATHEMATICS, Lucas N. H. Bunt, Phillip S. Jones, and Jack D. Bedient. Exciting, hands-on approach to understanding fundamental underpinnings of modern arithmetic, algebra, geometry and number systems examines their origins in early Egyptian, Babylonian, and Greek sources. 336pp. 5 3/8 x 8 1/2. 0-486-25563-8

THE THIRTEEN BOOKS OF EUCLID'S ELEMENTS, Euclid. Contains complete English text of all 13 books of the Elements plus critical apparatus analyzing each definition, postulate, and proposition in great detail. Covers textual and linguistic matters; mathematical analyses of Euclid's ideas; classical, medieval, Renaissance and modern commentators; refutations, supports, extrapolations, reinterpretations and historical notes. 995 figures. Total of 1,425pp. All books 5 3/8 x 8 1/2.

Vol. I: 443pp. 0-486-60088-2
Vol. II: 464pp. 0-486-60089-0
Vol. III: 546pp. 0-486-60090-4

A HISTORY OF GREEK MATHEMATICS, Sir Thomas Heath. This authoritative two-volume set that covers the essentials of mathematics and features every landmark innovation and every important figure, including Euclid, Apollonius, and others. 5 3/8 x 8 1/2.

Vol. I: 461pp. 0-486-24073-8
Vol. II: 597pp. 0-486-24074-6

A MANUAL OF GREEK MATHEMATICS, Sir Thomas L. Heath. This concise but thorough history encompasses the enduring contributions of the ancient Greek mathematicians whose works form the basis of most modern mathematics. Discusses Pythagorean arithmetic, Plato, Euclid, more. 1931 edition. 576pp. 5 3/8 x 8 1/2. 0-486-43231-9

CHINESE MATHEMATICS IN THE THIRTEENTH CENTURY, Ulrich Libbrecht. An exploration of the 13th-century mathematician Ch'in, this fascinating book combines what is known of the mathematician's life with a history of his only extant work, the Shu-shu chiu-chang. 1973 edition. 592pp. 5 3/8 x 8 1/2. 0-486-44619-0

PHILOSOPHY OF MATHEMATICS AND DEDUCTIVE STRUCTURE IN EUCLID'S ELEMENTS, Ian Mueller. This text provides an understanding of the classical Greek conception of mathematics as expressed in Euclid's Elements. It focuses on philosophical, foundational, and logical questions and features helpful appendixes. 400pp. 6 1/2 x 9 1/4. 0-486-45300-6

BEYOND GEOMETRY: Classic Papers from Riemann to Einstein, Edited with an Introduction and Notes by Peter Pesic. This is the only English-language collection of these 8 accessible essays. They trace seminal ideas about the foundations of geometry that led to Einstein's general theory of relativity. 224pp. 6 1/8 x 9 1/4. 0-486-45350-2

HISTORY OF MATHEMATICS, David E. Smith. Two-volume history – from Egyptian papyri and medieval maps to modern graphs and diagrams. Non-technical chronological survey with thousands of biographical notes, critical evaluations, and contemporary opinions on over 1,100 mathematicians. 5 3/8 x 8 1/2.

Vol. I: 618pp. 0-486-20429-4
Vol. II: 736pp. 0-486-20430-8

Physics

THEORETICAL NUCLEAR PHYSICS, John M. Blatt and Victor F. Weisskopf. An uncommonly clear and cogent investigation and correlation of key aspects of theoretical nuclear physics by leading experts: the nucleus, nuclear forces, nuclear spectroscopy, two-, three- and four-body problems, nuclear reactions, beta-decay and nuclear shell structure. 896pp. 5 3/8 x 8 1/2. 0-486-66827-4

QUANTUM THEORY, David Bohm. This advanced undergraduate-level text presents the quantum theory in terms of qualitative and imaginative concepts, followed by specific applications worked out in mathematical detail. 655pp. 5 3/8 x 8 1/2. 0-486-65969-0

ATOMIC PHYSICS AND HUMAN KNOWLEDGE, Niels Bohr. Articles and speeches by the Nobel Prize–winning physicist, dating from 1934 to 1958, offer philosophical explorations of the relevance of atomic physics to many areas of human endeavor. 1961 edition. 112pp. 5 3/8 x 8 1/2. 0-486-47928-5

COSMOLOGY, Hermann Bondi. A co-developer of the steady-state theory explores his conception of the expanding universe. This historic book was among the first to present cosmology as a separate branch of physics. 1961 edition. 192pp. 5 3/8 x 8 1/2. 0-486-47483-6

LECTURES ON QUANTUM MECHANICS, Paul A. M. Dirac. Four concise, brilliant lectures on mathematical methods in quantum mechanics from Nobel Prize-winning quantum pioneer build on idea of visualizing quantum theory through the use of classical mechanics. 96pp. 5 3/8 x 8 1/2. 0-486-41713-1

THE PRINCIPLE OF RELATIVITY, Albert Einstein and Frances A. Davis. Eleven papers that forged the general and special theories of relativity include seven papers by Einstein, two by Lorentz, and one each by Minkowski and Weyl. 1923 edition. 240pp. 5 3/8 x 8 1/2. 0-486-60081-5

PHYSICS OF WAVES, William C. Elmore and Mark A. Heald. Ideal as a classroom text or for individual study, this unique one-volume overview of classical wave theory covers wave phenomena of acoustics, optics, electromagnetic radiations, and more. 477pp. 5 3/8 x 8 1/2. 0-486-64926-1

THERMODYNAMICS, Enrico Fermi. In this classic of modern science, the Nobel Laureate presents a clear treatment of systems, the First and Second Laws of Thermodynamics, entropy, thermodynamic potentials, and much more. Calculus required. 160pp. 5 3/8 x 8 1/2. 0-486-60361-X

QUANTUM THEORY OF MANY-PARTICLE SYSTEMS, Alexander L. Fetter and John Dirk Walecka. Self-contained treatment of nonrelativistic many-particle systems discusses both formalism and applications in terms of ground-state (zero-temperature) formalism, finite-temperature formalism, canonical transformations, and applications to physical systems. 1971 edition. 640pp. 5 3/8 x 8 1/2. 0-486-42827-3

QUANTUM MECHANICS AND PATH INTEGRALS: Emended Edition, Richard P. Feynman and Albert R. Hibbs. Emended by Daniel F. Styer. The Nobel Prize–winning physicist presents unique insights into his theory and its applications. Feynman starts with fundamentals and advances to the perturbation method, quantum electrodynamics, and statistical mechanics. 1965 edition, emended in 2005. 384pp. 6 1/8 x 9 1/4. 0-486-47722-3

Physics

INTRODUCTION TO MODERN OPTICS, Grant R. Fowles. A complete basic undergraduate course in modern optics for students in physics, technology, and engineering. The first half deals with classical physical optics; the second, quantum nature of light. Solutions. 336pp. 5 3/8 x 8 1/2. 0-486-65957-7

THE QUANTUM THEORY OF RADIATION: Third Edition, W. Heitler. The first comprehensive treatment of quantum physics in any language, this classic introduction to basic theory remains highly recommended and widely used, both as a text and as a reference. 1954 edition. 464pp. 5 3/8 x 8 1/2. 0-486-64558-4

QUANTUM FIELD THEORY, Claude Itzykson and Jean-Bernard Zuber. This comprehensive text begins with the standard quantization of electrodynamics and perturbative renormalization, advancing to functional methods, relativistic bound states, broken symmetries, nonabelian gauge fields, and asymptotic behavior. 1980 edition. 752pp. 6 1/2 x 9 1/4. 0-486-44568-2

FOUNDATIONS OF POTENTIAL THERY, Oliver D. Kellogg. Introduction to fundamentals of potential functions covers the force of gravity, fields of force, potentials, harmonic functions, electric images and Green's function, sequences of harmonic functions, fundamental existence theorems, and much more. 400pp. 5 3/8 x 8 1/2. 0-486-60144-7

FUNDAMENTALS OF MATHEMATICAL PHYSICS, Edgar A. Kraut. Indispensable for students of modern physics, this text provides the necessary background in mathematics to study the concepts of electromagnetic theory and quantum mechanics. 1967 edition. 480pp. 6 1/2 x 9 1/4. 0-486-45809-1

GEOMETRY AND LIGHT: The Science of Invisibility, Ulf Leonhardt and Thomas Philbin. Suitable for advanced undergraduate and graduate students of engineering, physics, and mathematics and scientific researchers of all types, this is the first authoritative text on invisibility and the science behind it. More than 100 full-color illustrations, plus exercises with solutions. 2010 edition. 288pp. 7 x 9 1/4. 0-486-47693-6

QUANTUM MECHANICS: New Approaches to Selected Topics, Harry J. Lipkin. Acclaimed as "excellent" (*Nature*) and "very original and refreshing" (*Physics Today*), these studies examine the Mössbauer effect, many-body quantum mechanics, scattering theory, Feynman diagrams, and relativistic quantum mechanics. 1973 edition. 480pp. 5 3/8 x 8 1/2. 0-486-45893-8

THEORY OF HEAT, James Clerk Maxwell. This classic sets forth the fundamentals of thermodynamics and kinetic theory simply enough to be understood by beginners, yet with enough subtlety to appeal to more advanced readers, too. 352pp. 5 3/8 x 8 1/2. 0-486-41735-2

QUANTUM MECHANICS, Albert Messiah. Subjects include formalism and its interpretation, analysis of simple systems, symmetries and invariance, methods of approximation, elements of relativistic quantum mechanics, much more. "Strongly recommended." – *American Journal of Physics.* 1152pp. 5 3/8 x 8 1/2. 0-486-40924-4

RELATIVISTIC QUANTUM FIELDS, Charles Nash. This graduate-level text contains techniques for performing calculations in quantum field theory. It focuses chiefly on the dimensional method and the renormalization group methods. Additional topics include functional integration and differentiation. 1978 edition. 240pp. 5 3/8 x 8 1/2. 0-486-47752-5

Physics

MATHEMATICAL TOOLS FOR PHYSICS, James Nearing. Encouraging students' development of intuition, this original work begins with a review of basic mathematics and advances to infinite series, complex algebra, differential equations, Fourier series, and more. 2010 edition. 496pp. 6 1/8 x 9 1/4. 0-486-48212-X

TREATISE ON THERMODYNAMICS, Max Planck. Great classic, still one of the best introductions to thermodynamics. Fundamentals, first and second principles of thermodynamics, applications to special states of equilibrium, more. Numerous worked examples. 1917 edition. 297pp. 5 3/8 x 8. 0-486-66371-X

AN INTRODUCTION TO RELATIVISTIC QUANTUM FIELD THEORY, Silvan S. Schweber. Complete, systematic, and self-contained, this text introduces modern quantum field theory. "Combines thorough knowledge with a high degree of didactic ability and a delightful style." – *Mathematical Reviews.* 1961 edition. 928pp. 5 3/8 x 8 1/2. 0-486-44228-4

THE ELECTROMAGNETIC FIELD, Albert Shadowitz. Comprehensive undergraduate text covers basics of electric and magnetic fields, building up to electromagnetic theory. Related topics include relativity theory. Over 900 problems, some with solutions. 1975 edition. 768pp. 5 5/8 x 8 1/4. 0-486-65660-8

THE PRINCIPLES OF STATISTICAL MECHANICS, Richard C. Tolman. Definitive treatise offers a concise exposition of classical statistical mechanics and a thorough elucidation of quantum statistical mechanics, plus applications of statistical mechanics to thermodynamic behavior. 1930 edition. 704pp. 5 5/8 x 8 1/4. 0-486-63896-0

INTRODUCTION TO THE PHYSICS OF FLUIDS AND SOLIDS, James S. Trefil. This interesting, informative survey by a well-known science author ranges from classical physics and geophysical topics, from the rings of Saturn and the rotation of the galaxy to underground nuclear tests. 1975 edition. 320pp. 5 3/8 x 8 1/2. 0-486-47437-2

STATISTICAL PHYSICS, Gregory H. Wannier. Classic text combines thermodynamics, statistical mechanics, and kinetic theory in one unified presentation. Topics include equilibrium statistics of special systems, kinetic theory, transport coefficients, and fluctuations. Problems with solutions. 1966 edition. 532pp. 5 3/8 x 8 1/2. 0-486-65401-X

SPACE, TIME, MATTER, Hermann Weyl. Excellent introduction probes deeply into Euclidean space, Riemann's space, Einstein's general relativity, gravitational waves and energy, and laws of conservation. "A classic of physics." – *British Journal for Philosophy and Science.* 330pp. 5 3/8 x 8 1/2. 0-486-60267-2

RANDOM VIBRATIONS: Theory and Practice, Paul H. Wirsching, Thomas L. Paez and Keith Ortiz. Comprehensive text and reference covers topics in probability, statistics, and random processes, plus methods for analyzing and controlling random vibrations. Suitable for graduate students and mechanical, structural, and aerospace engineers. 1995 edition. 464pp. 5 3/8 x 8 1/2. 0-486-45015-5

PHYSICS OF SHOCK WAVES AND HIGH-TEMPERATURE HYDRO DYNAMIC PHENOMENA, Ya B. Zel'dovich and Yu P. Raizer. Physical, chemical processes in gases at high temperatures are focus of outstanding text, which combines material from gas dynamics, shock-wave theory, thermodynamics and statistical physics, other fields. 284 illustrations. 1966–1967 edition. 944pp. 6 1/8 x 9 1/4. 0-486-42002-7